INTRODUCTION TO NONLINEAR DYNAMICS
FOR PHYSICISTS

World Scientific Lecture Notes in Physics – Vol. 53

INTRODUCTION TO NONLINEAR DYNAMICS FOR PHYSICISTS

Henry D. I. Abarbanel
M. I. Rabinovich
M. M. Sushchik

Department of Physics and
 Institute for Nonlinear Science
University of California, San Diego
La Jolla, Ca 92093-0402
USA

World Scientific
Singapore • New Jersey • London • Hong Kong

Published by

World Scientific Publishing Co. Pte. Ltd.

P O Box 128, Farrer Road, Singapore 912805

USA office: Suite 1B, 1060 Main Street, River Edge, NJ 07661

UK office: 57 Shelton Street, Covent Garden, London WC2H 9HE

British Library Cataloguing-in-Publication Data
A catalogue record for this book is available from the British Library.

First published 1993
First reprint 1996

ISBN 981-02-1409-X
     981-02-1410-3 (pbk)

Printed in Singapore.

Preface

This is a series of lectures on nonlinear dynamics for physicists. The level is that of an advanced undergraduate or beginning graduate student. The assumption is made that the student has completed an advanced mechanics course and has had some exposure to mathematical physics of the sort encountered in an advanced electricity and magnetism or quantum mechanics course. No detailed knowledge of nonlinear dynamics is assumed, but familiarity with ordinary differential equations is. Some of the material is cast at a level which is more challenging to persons having only an undergraduate background in Physics but we felt that the subject was sufficiently interesting and rich that posing this kind of challenge was within the scope of the course. Similarly some of the material is aimed below that of a typical first year graduate student, but we also felt that this would serve both as a reminder and as a presentation of perhaps a different point of view for such students.

In any case the main goal of the lectures is to present both substantial qualitative information about phenomena in nonlinear systems to the advanced Physics student making the flavor tantalizing, yet providing sufficient quantitative material that the student would learn how to proceed in similar situations without the instructors "magic wand", which seems, we admit, to be waved here and there. Thus we set ourselves an impossible challenge overall, but we hope that the reader will find some success in our achieving these goals. Anyway, we trust enough success that we can hope for feedback from readers on topics (both those included and those left out), difficulty of presentation, and other items related to scientific material and pedagogy.

We hope that the reader, and even more the person who both hears the lectures and read these notes as they evolve to more finished stages, will have addressed three main questions about nonlinear dynamics:

- What is nonlinear dynamics all about and what makes it differ from linear dynamics which permeates all familiar textbooks?

- From the physicist's point of view, why should we study nonlinear systems and leave the comfortable territory of linearity?

- How can one progress in the study of nonlinear systems both in the analysis of these systems when we know them and in learning about new systems from observations of their experimental behavior?

Of course, neither these lectures nor whole lifetimes of work can possibly answer all of these questions in the finest detail. Nonetheless, we have set these impossible goals so that the student will reach with us slightly beyond an easy grasp.

These lectures were originally given by M. I. Rabinovich as Physics 155 in the UCSD Physics Department during the Fall Quarter of the 1992/93 academic year. Notes were taken and then transcribed by M. M. Sushchik, and edited by H. D. I. Abarbanel. It is our plan that this course will be taught by Abarbanel in the 1993/94 academic year at UCSD, and that they will be improved based on that experience and extensive input from the students of each course. Eventually we are thinking of making these notes into a monograph/textbook, but the barriers for turning a useful pedagogical product into a polished book are known to be high.

The problems were not part of the course when it was first taught, but they have been added by us as a guide for the reader, and we do plan to incorporate them in subsequent presentations of the material. We actively solicit feedback from readers of these notes. They are meant to be unpolished, yet, hopefully, transmit substantial material which is of interest to Physicists.

Henry D. I. Abarbanel
M. I. Rabinovich
M. M. Sushchik
January, 1993

Contents

INTRODUCTION TO NONLINEAR DYNAMICS FOR PHYSICISTS

1 Introduction

What is dynamics? What is **nonlinear** dynamics? Let us start with the simple example of the swing on a child's playground. The child is raising and lowering herself by pumping her legs back and forth. Up and down, up and down. The swing oscillates up and down, and with strong pumping could even turn upside-down and start rotating.

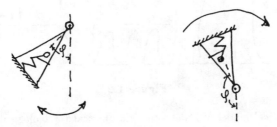

Figure 1.1

What phenomena can we see in this motion?

First of all we see ordinary and bounded oscillations, and we may also see rotations completely around the pivot of the swing if the child is vigorous enough. What are the energy sources for this motion? Of course, our little guy has a lot of energy. In this example we will see how the energy of our guy pumping the swing is transformed through an **instability** (the so-called parametric instability) into the energy of oscillations. One might think that the behavior of the swing is simple if the guy's behavior is simple. **THIS IS NOT QUITE TRUE!!!** The pendulum behavior may be very complex even when the behavior of our guy is very simple. This means that the pendulum has its own behavior which is rather independent of the guy or

1

any other specific source of pumping or energy transfer to the swing. The study of this kind of behavior is what we call dynamics.

How can we describe this set of phenomena?

We are physicists, so we turn to a simple mathematical model in which to think about our little guy and her swing. If the properties of our guy's behavior are independent of the properties of the swing; that is, there is no feedback from the swing to the guy and the guy remains the same while she swings (except, of course, she is having a lot of fun), then we can write the following equation to describe the situation:

$$\ddot{\phi}(t) + h\dot{\phi}(t) + \frac{g}{l(t)} \sin[\phi(t)] = 0. \tag{1}$$

Here ϕ is the angle of the swing with respect to the direction of gravity [vertical] (see Figure 1.1). $l(t)$ is the effective length of the swing. g is the gravitational acceleration. h is a friction coefficient, and the mass of the guy is set to unity.

When our guy is tired and decides to have a rest $l(t) = $ const, and this equation coincides with the usual (damped) pendulum equation.

If we plot the angle as a function of time while our guy is pumping, we will see something like this:

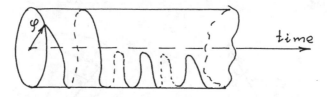

Figure 1.2

A mathematical model of a deterministic physical (or biological or chemical or any other) system is represented by a dynamical system.

Our Definition of a Dynamical System: A **DYNAMICAL SYSTEM** is a mathematical description corresponding to a real system whose evolution depends uniquely on the initial state.

A dynamical system is described by a system of equations (differential, difference, integral, etc.) which allows for the existence of a unique finite (i.e. bounded) solution for each initial condition over an infinite period of time.

For example, the pendulum equation

$$\ddot{\phi} + h\dot{\phi} + \omega^2(t) \sin \phi = 0 \tag{2}$$

describes a dynamical system, if ω is a deterministic function of time. If $\omega(t)$ is a random function of time, then everything changes and from any initial condition can come an infinite number of possible trajectories. The collection of them has a probability distribution.

The system described by

$$da/dt = a^2 \tag{3}$$

is not an acceptable dynamical system from the physicists point of view. The mathematical solution is

$$a(t) = 1/(1/a(0) - t). \tag{4}$$

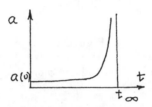

Figure 1.3

When t goes to $t_\infty = \frac{1}{a(0)}$ the solution $a(t)$ goes to infinity. Therefore, we know nothing about the behavior of the solution at times greater than t_∞. This phenomena is called an explosive instability, and it indicates that we have a quite limited model from the point of view of physics.

A state of a dynamical system is described by a set of variables (angle, velocity, voltage, etc.). There are different criteria for choosing a particular set of variables for the physical description: symmetry and/or simplicity considerations, natural interpretation, etc. We call the set of variables required to describe the physics of the system a **phase space**. In the case of the pendulum pumped by our guy the phase space is three dimensional and has useful, physical coordinates made out of $\phi(t)$, $\dot{\phi}(t)$, and t.

Each state corresponds to a point in the phase space while evolution is depicted by trajectories moving among these points in phase space. The location of states relative to each other in the phase space of a dynamical system is described by the notion of distance. A collection of states at a fixed moment of time fills a phase volume of the system. The behavior of phase trajectories depicts qualitatively the evolution of a dynamical system. For example, a degenerate trajectory (just a point in a phase space) corresponds to an equilibrium or time independent or fixed point state, and a closed trajectory represents a periodic motion.

A dynamical system is given the law by which each point is related to a point on the phase trajectory at the previous moments of time. For example, for the equation $\ddot{\phi} + \omega^2 \sin(\phi) = 0$ the phase space looks like this:

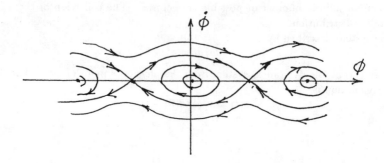

Figure 1.4

Noting that the phase space is periodic in ϕ one can bend it into a cylinder.

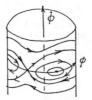

Figure 1.5

As you probably remember, in a linear system we can see only one equilibrium state and only one type of behavior. In our **NONLINEAR** system, however, we can see different equilibrium points and different kinds of motions. The line which separates qualitatively different kinds of motions is referred to as a separatrix. In the phase space we have just shown of the undamped pendulum, the separatrix is the curve which connects the fixed points (the **X** points in the Figure) at $\phi = 0$ or $\phi = 2\pi$ while $\dot{\phi} = 0$.

Let us go back to the swing. Please recall your own experience. You (or our guy) are pumping up or down with respect to the seat DEPENDING ON THE POSITION OF THE SWING AND ITS AMPLITUDE. This is a very important element of nonlinear systems called feedback. If you want to succeed in pumping the swing up, you cannot raise and lower yourselves periodically! This is because the swing is a nonlinear system, and therefore the period of its oscillations depends on the amplitude.

What if the guy is not as smart as we would hope, and his behavior is completely periodic ($l(t) = l_0[1 + b\sin\Omega t]$)? When $b << 1$, it does not matter too much. But if $b \sim 1$ the oscillations of the swing will be chaotic. It is quite likely that one of the parents of the small guy will run to rescue her long before this!

Our experience and intuition as well as traditional education indicates that in a dynamical system described by a regular law (equations) nothing irregular or random or stochastic should occur. Where does randomness come from? We see that in the case of the swing, the motion near the separatrix is very sensitive to any perturbation. Our guy can drive the system from the region of rotations around the pivot of the swing to the region of bounded oscillations and back, and she can do all this by small perturbations near the separatrix. Changing back and forth between the two regimes may lead to a very irregular looking motion in our very simple looking system. This irregular motion we call chaos, and we will learn much about it in this course.

Our chaos appears in a law governed by an unambiguous algorithm that uniquely defines for a given initial condition the behavior of the system in its future! Obviously, for very complex systems (systems which have large numbers of degrees of freedom, such as a gas in a vessel) the dynamics (behavior) is so complex that we need a fully statistical description above and beyond the chaos surely also lurking in the dynamics.

2 Nonlinear Oscillator without Dissipation

1. A ball in a trough

$$
\begin{aligned}
m\ddot{x} &= -F \\
F &= mg\sin\phi = mg\,dz/dx \\
\ddot{x} + g\,dz/dx &= 0 \\
\text{or } \ddot{x} &= -\frac{\partial W}{\partial x}
\end{aligned}
\tag{5}
$$

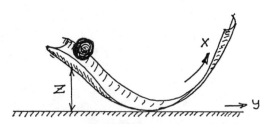

Figure 2.1

$W(x) = gz(x)$ is called the potential energy or potential. When $W(x) = x^2$ (the potential well is parabolic) our oscillator is linear, and the equation of motion becomes $\ddot{x} + \omega^2 x = 0$.

One important point is that the shape of the potential does not coincide with that of the trough in the (z, y)-plane. If, for example, the trough equation is $z = y^2$, then

6

$dy = \frac{1}{2}dz/\sqrt{z}$. We have $dx^2 = dz^2 + dy^2$, $dx^2 = dz^2[1 + \frac{1}{4z}]$ and

$$x = \int dz \sqrt{(1 + \frac{1}{4z})}.$$ (6)

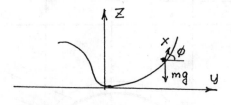

Figure 2.2

This is not a parabolic potential $x \sim \sqrt{z}$!

2. Our equations of motion can be easily integrated:

$$dx/dt = v;\ dv/dt = -\partial W/\partial x \implies (dv/dt)/(dx/dt) = -(\partial W/\partial x)/v$$

$$dv/dx = -(1/v)\cdot(\partial W/\partial x) \implies v^2/2 + W(x) = E = \text{const}$$

E is the total energy of the nonlinear oscillator.

$$E = W_{\text{kin}} + W_{\text{pot}} = (dx/dt)^2/2 + W(x)$$

A qualitative picture of the motion of an energy conserving nonlinear oscillator with one degree of freedom, as here, can be represented without solving the differential equations simply in terms of its phase plane (that is, the coordinates (x, v)). Using the conservation of energy as expressed above, we can write the velocity as

$$v = dx/dt = \pm\sqrt{2\cdot[E - W(x)]}$$

This is in fact the equation for trajectories in the phase plane of our nonlinear oscillator. For different values of E we have different trajectories.

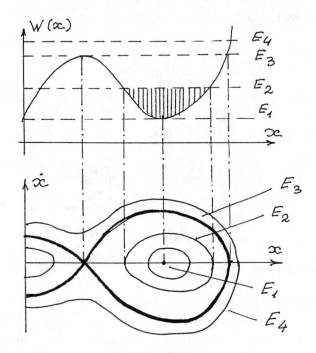

Figure 2.3

For example, if $W = x^2/c$ we have the equation for an ellipse:

$$v = \pm\sqrt{2 \cdot [E - x^2/c]} \text{ or } v^2/2 + x^2/c = \text{const}$$

This means that harmonic oscillations $\ddot{x} + \omega^2 \cdot x = 0$ appear as ellipses in the phase plane.

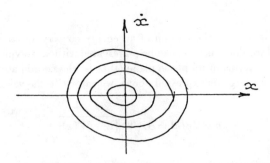

Figure 2.4

If we plot $W(x)$, we will notice that there are points where both the velocity and the acceleration are zero.

$$\dot{x} = v = 0$$

$$\dot{v} = -\partial W/\partial x = 0$$

We will call those points equilibrium points or equilibrium states. The second equation says that the equilibrium points are the points where $\partial W/\partial x = 0$; that is, the points where the force on the ball vanishes. It is quite clear that if you place a ball on the top of a hill, it will stay there forever unless you push it. The same will happen if you put the ball in a pit.

3. The integral curves on which the direction of motion is defined are referred to as phase trajectories.

As noted above, phase trajectories depict the evolution of our dynamical system in time. We can determine the direction of motion along integral curves looking at the velocity equation $dx/dt = v$. In our case the arrows are directed to the right when $v > 0$ and to the left when $v < 0$.

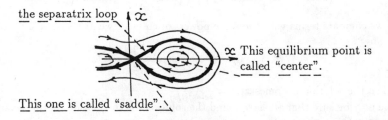

Figure 2.5

The separatrix loop corresponds to a pulse with infinitely long "tails". This is because the system has zero velocity near the saddle point and the time it takes to reach this point is logarithmically infinite. If we consider a trajectory close to a separatrix, the motion will be a sequence of pulses:

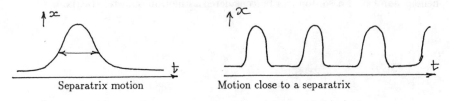

Figure 2.6

4. Analytical description of motion on a separatrix loop

Example:
$$W(x) = -\frac{x^2}{2} + \frac{x^3}{3}$$
$$\ddot{x} - x + x^2 = 0$$

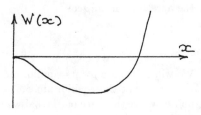

Figure 2.7

The solution to this equation is given by

$$x(t) = x_0/\cosh^2(t/\tau)$$

where τ is the characteristic width of the pulse.

This solution can be easily verified by direct substitution:

$$\frac{1}{\tau^2}\left\{ \frac{4x_0}{\cosh^2\left(\frac{t}{\tau}\right)} - \frac{6x_0}{\cosh^4\left(\frac{t}{\tau}\right)} \right\} - \frac{x_0}{\cosh^2\left(\frac{t}{\tau}\right)} + \frac{x_0^2}{\cosh^4\left(\frac{t}{\tau}\right)} \equiv 0$$

If we compare terms with the same powers of $\cosh(\bullet)$, we find values for τ and x_0

$$4/\tau^2 = 1; \ 6/\tau^2 = x_0 => x_0 = \frac{3}{2}; \ \tau = 2.$$

Then the solution becomes $x(t) = \frac{3}{2\cosh^2(t/2)}$.

We may be sure that we have found the solution for the motion on the separatrix because in this system for $x > 0$ there is only one solution with infinite period. Suppose we start at some time and move ahead in time along the orbit $x(t)$. At time t_1 we will have reached $x(t_1)$, which, if $t_1 \gg 2$ is approximately

$$x(t_1) \approx 6\exp[-t_1], \tag{7}$$

so the time required to reach $x(\infty) = 0$ goes as $\log[x(\infty)]$.

This kind of motion is very important for studying nonlinear waves because the phenomenon called a soliton can be considered as motion on a separatrix.

3 Equilibrium States of a Nonlinear Oscillator with Dissipation

Where dissipation appears, the energy conservation law disappears. But different kinds of behavior change differently when dissipation is turned on. For example close, to a center, where a nondissipative (also called Hamiltonian) system is stable, the behavior will change substantially, while near a saddle point, where the Hamiltonian system is unstable, there will be just small changes. This is clear intuitively. In both cases (with or without friction) a ball will roll down from a hill (a saddlepoint). But, if we have friction, oscillations (around a center) will decay, and we will not see periodic motion.

Now let us study in some detail how a dissipative system behaves near an equilibrium point. Any system with **two degrees of freedom** may be described by the set of equations:

$$\begin{aligned} dx/dt &= f(x,y) \\ dy/dt &= g(x,y). \end{aligned} \tag{8}$$

Near an equilibrium, (x_0, y_0), namely a time independent solution to these equations, we can describe our system by a set of *linearized* equations. Since at equilibrium $f(x_0, y_0) = g(x_0, y_0) = 0$ these equations may be written in the form

$$\begin{aligned} \frac{dX}{dt} &= \frac{df}{dx}(x_0, y_0)X + \frac{df}{dy}(x_0, y_0)Y \\ \frac{dY}{dt} &= \frac{dg}{dx}(x_0, y_0)X + \frac{dg}{dy}(x_0, y_0)Y. \end{aligned} \tag{9}$$

Here we introduced $X(t) = x(t) - x_0$ and $Y(t) = y(t) - y_0$, which are assumed to be small $(|X(t)|, |Y(t)| \ll 1)^2$. Through linearization we have reduced the problem to

---

[2]Here we do not consider the case when the linear terms in Taylor expansion are zero, as in the case of $d^2x/dt^2 + x^3 = 0$. In such cases one may also carry out this discussion, but it is the usual case that the leading linear terms will dominate.

a very simple one:

$$
\begin{aligned}
dX/dt &= aX + bY \\
dY/dt &= cX + dY
\end{aligned}
\tag{10}
$$

We know from linear algebra that the solution will have the form

$$
\begin{aligned}
X(t) &= X_0 e^{\lambda t} \\
Y(t) &= Y_0 e^{\lambda t},
\end{aligned}
\tag{11}
$$

and that the condition for nonzero solutions is $\det |A - \lambda E| = 0$. (E is the identity matrix, and A is the two by two matrix of coefficients in the equation for $X(t)$ and $Y(t)$.)

The solutions λ of this equation are called the characteristic exponents. Let us see what kind of information they can provide us with in the simple example of a linear oscillator with friction:

$$
d^2x/dt^2 + 2\mu dx/dt + \Omega^2 x = 0
$$

We can rewrite this in the form of a set of first order equations. This is a standard method in nonlinear dynamics since it identifies the full number of degrees of freedom we need to consider:

$$
dx/dt = y
$$

$$
dy/dt = -2\mu y - \Omega^2 x.
$$

For this system

$$
\det |A - \lambda E| = \lambda^2 + 2\mu\lambda + \Omega^2
$$

The characteristic exponents are

$$
\lambda_{1,2} = -\mu \pm \sqrt{\mu^2 - \Omega^2}
$$

We can see that the plane of parameters is broken into five regions with qualitatively different behaviors. The behavior in a particular region is determined by the positions of $\lambda_{1,2}$ in the complex plane. Figure 3.1 shows the partitioning of the parameter plane, the positions of $\lambda_{1,2}$ in the complex plane in each region and the behavior of the system near the stationary or fixed point or equilibrium point $dx/dt = x = 0$.[3]

---

[3]This is a trivial example of partitioning of the parameter plane into regions with qualitatively different behavior. In a more interesting case we may have a richer variety of behavior: oscillatory behavior, quasi-periodic behavior, chaotic behavior and so on.

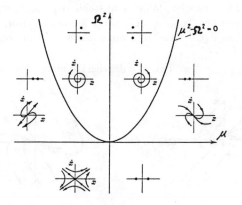

Figure 3.1

The various kinds of behavior are categorized as follows:

(1) If $\mu = 0$, that is no friction, $\lambda_{1,2} = \pm i\Omega$. This equilibrium state is called a center, and we referred to it above.

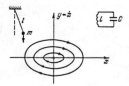

Figure 3.2

We have seen this picture before, when we studied systems without dissipation.

(2) If $\Omega > \mu > 0$, $\lambda_{1,2} = -\mu \pm i\sqrt{\Omega^2 - \mu^2}$. This equilibrium state is called a focus.

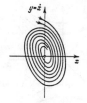
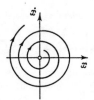

Figure 3.3

As you see a focus can be stable or unstable depending on the sign of μ. A stable focus describes damped oscillations of the oscillator. An unstable focus describes the "no signal state" of a generator with "soft" onset of oscillations.

(3) If $\Omega = \mu$, then $\lambda_{1,2} = -\mu$. This is a degenerate node.

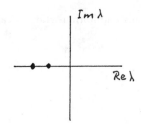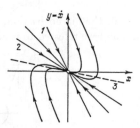

Figure 3.4

(4) When $\Omega < \mu, \lambda_{1,2} = -\mu \pm \sqrt{\mu^2 - \Omega^2}$. This point is called a node.

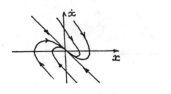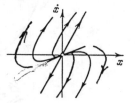

Figure 3.5

(5) If $\Omega^2 < 0, \lambda_{1,2} = -\mu \pm \sqrt{\mu^2 - |\Omega^2|}$. This is already familiar to us as a saddle. This is a very important type of equilibrium state which underlies the existence of both solitons and chaos.

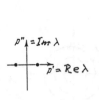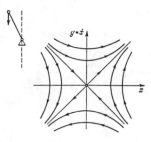

Figure 3.6

Now assume that the dissipation is very large compared to the frequency. Then λ_1 is very close to zero, and $\lambda_2 = -2 \cdot \mu$. We have a node but with very different time scales for motions along different directions. Our system comes very quickly on the line $dx/dt = cx$ (where $c \approx 0$), and then the subsequent behavior of the system is on the phase line and not throughout the phase plane. This is a very important conclusion.

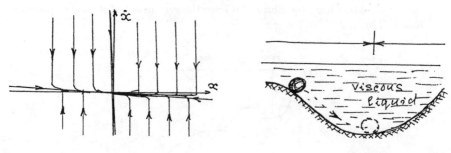

Figure 3.7	Figure 3.8

An example of such a system is a ball in a very viscous liquid. We can assume that the acceleration is very small, and we can write the equation of motion balancing gravity forces and viscous forces.

In a general case we can write the equation for motion on a phase line as:

$$dA/dt = F(A),$$

and it is easy to find the solution:

$$dA/F(A) = dt \rightarrow \int dA'/F(A') = t - t_0$$

We can rewrite the equation of motion in the form[4]

$$\frac{dA}{dt} = -\frac{\partial U}{\partial A} \quad \text{with} \quad U[A] = -\int F(A')dA'.$$

$U[A]$ is a potential (sometimes also called the free energy).

For the time derivative of $U[A]$ we can write the following equation:

$$dU/dt = (dU/dA)(dA/dt) = -(dA/dt)^2 \leq 0.$$

---

[4]Systems described by this equation are called gradient systems. They are very important, and we will come back to them when we start considering processes of pattern formation.

This is a very important result which is pretty obvious, if you use the analogy with a ball in a viscous liquid. The energy of the system (or of the ball) can only decrease. Therefore, if we release our system from some state it will try to find the minimum of $U(A)$, and after that it will stay there forever. It is quite obvious that a maximum of $U(A)$ is an unstable point. But also we can see one more type of stationary point where both first and second derivatives of $U(A)$ are zero and but the third derivative is not. In this case we need to introduce a larger deviation from equilibrium to force the system out of it than in previous cases. This state is called semistable.

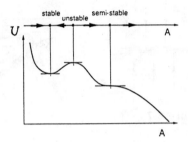

Figure 3.9

4 Oscillations in Systems with Nonlinear Dissipation–Generators

By now we have gotten acquainted with three types of oscillations:

(1) undamped oscillations in a nonlinear systems without dissipation (conservative or Hamiltonian systems);

(2) damped oscillations in nonlinear systems with linear dissipation;

(3) growing oscillations in a nonlinear oscillator with negative dissipation.

But both in natural settings such as large scale weather dynamics and in laboratory settings we can see many examples of undamped oscillations in nonlinear systems. We call systems which allow for such oscillations **generators**; this term is used widely in the Russian literature and rarely in the American or European or Japanese literature, but we are introducing it here. Here are a few simple examples: the oscillations of a violin string when the bow is moved across the string; an ordinary AM or FM radio; the pressure and density (acoustic) oscillations of air in an organ pipe; the oscillations of the pendulum in a clock mechanism.

 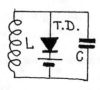 

Figure 4.1

In these and many other cases, undamped oscillation arises as the result of a basic instability which is stabilized via nonlinear damping.

Let us study the following mathematical model which describes a wide variety of such systems:

$$d^2x/dt^2 + x = \mu[-q(x)dx/dt + f(x)]$$

where $\mu << 1$, or equivalently,

$$dx/dt = y$$

$$dy/dt = -x + \mu[-q(x)y + f(x)],$$

writing our equation in the standard first order form.

It is very convenient to introduce a new variable $a(t)$ (which is complex and therefore contains two independent pieces of information $Re(a)$ and $Im(a)$). This variable is defined as follows

$$x(t) = a(t) \cdot \exp(it) + a^*(t) \cdot \exp(-it)$$

$$y(t) = i \cdot a(t) \cdot \exp(it) - i \cdot a^* \cdot \exp(-it),$$

and is similar to the introduction of annihilation and creation operators in quantum theory.

After we substitute these expressions for x and y in terms of $a(t)$ into the original equations we arrive at

$$da/dt \exp(it) + da^*/dt \exp(-it) = 0$$

$$ida/dt \exp(it) - ida^*/dt \exp(-it) =$$

$$\mu[-q(a, a^*, \exp(\pm it))(ia(t) \exp(it) - ia^* \exp(\pm it)) +$$

$$f(a, a^*, \exp(\pm it))].$$

Using the first equation we can rewrite the second one in the form

$$da/dt = [\mu/(2i)][-q(a, a^*, \exp(\pm it))]$$
$$(ia \exp(it) - ia^* \exp(-it)) + f(a, a^*, \exp(\pm it))] \exp(-it). \qquad (12)$$

At this point we have not made any approximations yet, and the last equation is identical to our starting point.

Now we consider the case when $\mu << 1$. Then we notice that $da/dt << a$. This is very reasonable: if the nonlinearity and dissipation are small, the oscillations are almost harmonic — that is da/dt is very close to zero. That is, $a(t)$ does not change much over the period of oscillations.

Then we can write our equation in the form

$$da/dt \approx [\mu/(2i)] \sum_{n=0}^{\infty} F^n(a, a^*) \exp(int)$$

Note that if $q(x)$ and $f(x)$ are polynomials, we will have $\sum_{n=0}^{M}$ instead of $\sum_{n=0}^{\infty}$ in this formula. M is the order of the polynomial. The Fourier coefficients F^n are defined by

$$F^n = (1/T) \int_{t}^{t+T} [-q(a, a^*, \exp(\pm it))(ia \exp(it) - ia \exp(-it) +$$
$$f(a, a^*, \exp(\pm it))] \exp(-it) \exp(-int) dt.$$

This expansion of the righthand side is exact, if $a = $ const, and we assume that it holds approximately in the case when a changes very slowly. When we evaluate the integral we can assume $a(\mu t)$ to be a constant, since it does not change much over the time T. If we want to follow just the slow changes of a we need to keep only one term of this expansion, and that is the one with $n = 0$. This is obvious: we are writing an equation for a slow variable, hence this equation should not have fast time in it. Furthermore, we can prove mathematically that $\mu F^n \sim \mu^n$. Therefore the solution we obtain after a truncation at $n = 0$ will be correct to order μ^2. This is an example of perturbation theory for nonlinear oscillator.

Now let us consider the following concrete example:

$$q(x) = 1 - \alpha x^2 + \beta x^4, \ f(x) = 0$$

$$q(a, a^*, \exp(\pm it)) = 1-$$

$$\alpha(2|a|^2 + a^2 \exp(i \cdot 2t) + a^{*2} \exp(-i2t)) +$$

$$\beta(6|a|^4 + 4|a|(a \exp(it) + a^* \exp(-it)) +$$

$$(a^4 \exp(i4t) + a^{*4} \exp(-i4t)))$$

Using the truncation at $n = 0$ we find

$$F_0 = -(1 - 2\alpha|a|^2 + 6\beta|a|^4)ia,$$

and the equation for the slowly varying amplitude becomes

$$da/dt = -\mu[1 - 2\alpha|a|^2 + 6\beta|a|^4]a.$$

Let us now introduce new variables A, ϕ, T in the following way:

$$a = A \exp(i\phi); \ T = \mu t$$

$$dA/dT \exp(i\phi) + id\phi/dT \exp(i\phi) =$$

$$-[1 - 2\alpha A^2 + 6\beta A^4]A \exp(i\phi).$$

The equation for A will be

$$dA/dT = -\frac{\partial U[A]}{\partial A}$$

with $U[A] = A^2/2 - \alpha A^4/2 + \beta A^6$.

At first glance there is something confusing in this result: originally we had two equations of first order, and now we have just one. How did we manage to reduce the dimensionality of our phase space? Formally, we answer this question by writing the equation for the phase ϕ : $d\phi/dT = 0$. This implies we have oscillations at constant phase, and thus one of our degrees of freedom is "frozen" out in the $\mu \ll 1$ approximation.

To illustrate the physical meaning of this transition from two-dimensional phase space to one dimensional phase space, let us look at the trajectory in the original phase plane.

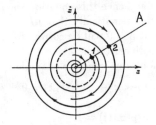

Figure 4.2

We can consider the amplitude at the points where the trajectory intersects a radial line. The amplitude changes very slowly between two intersections; that is, it changes quasicontinuously. The equation for A we just derived may be thought of as of an equation for those discrete A's written in the quasicontinuous limit.

The amplitude equation has the gradient form with potential $U[A]$ plotted below:

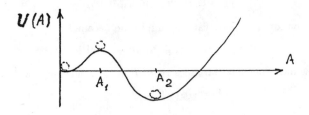

Figure 4.3

We have three fixed points. Two of them are stable, and one is unstable.

What does this mean in our original phase space? We considered only the equation for $a(t)$ but $x = a \exp(it) +$ cc, and, therefore, to reconstruct the solution in the phase plane we must rotate the solution for the amplitude equation at an angular frequency $\omega = 1$. Then we will see that the stationary points of the amplitude equation correspond to periodic motions. Unstable (stable) periodic trajectories are referred to as unstable (stable) <u>limit cycles</u>.

5 The Van der Pol Generator

As an example of a system where a limit cycle may occur let us consider the following circuit:

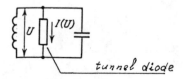

Figure 5.1

A tunnel diode is a nonlinear element which has the current-voltage characteristic shown in Figure 5.2.

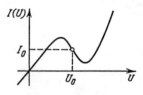

Figure 5.2

$$I = I_0 - g(U - U_0) + \beta'(U - U_0)^3$$

Tunnel diodes behave like an ordinary resistor at low and high voltages, but there is an intermediate region where it has negative resistance. We might expect that it would lead to amplification of small oscillations in the system (if we choose the parameters in a proper way) as a positive resistance leads to decaying oscillations.

It is easy to write the equations of motion for this system. We just need to know that

$$I_l + I(U) + I_c = 0$$

where

$$I_l = \frac{1}{L} \int U \, dt$$

and

$$I_c = C \frac{dU}{dt}$$

Then we can write

$$C \frac{dU}{dt} + I(U) + \frac{1}{L} \int U \, dt = 0$$

Differentiating this equation we arrive at

$$\frac{d^2U}{dt^2} + \frac{1}{C} \frac{dI(U)}{dt} + \frac{1}{LC} U = 0$$

or

$$\frac{d^2U}{dt^2} + \frac{1}{C} \left[-g + 3\,\beta'(U - U_0)^2 \right] \frac{dU}{dt} + \omega_0^2 U = 0$$

where we introduced the usual LC circuit resonant frequency

$$\omega_0^2 = \frac{1}{LC}.$$

It is very convenient to switch to the new variable

$$x = \frac{U - U_0}{U_0},$$

for which we will have the equation

$$\ddot{x} - \mu'(1 - \beta x^2)\dot{x} + \omega_0^2 x = 0$$

with

$$\mu' = \frac{g}{C}$$

and

$$\beta = \frac{CU_0^2}{g} 3\beta'$$

If we introduce a new time variable, we can write the equation in the form

$$\ddot{x} - \mu(1 - \beta x^2)\dot{x} + x = 0$$

where

$$\mu = \frac{\mu'}{\omega_0}.$$

This equation for an oscillator with nonlinear dissipation is sometimes called Van der Pol equation. One can see that $x = 0$ is an unstable point. This means that if the potential across the diode is in the region where the diode acts like a negative resistor, small deviations from this point will grow with time: exactly as we expected from physical considerations. The energy of the battery is thus transformed into the energy of oscillations.

Now let us consider the case of $\mu \ll 1$ and $\beta \approx 1$. We will apply the method discussed in the previous lecture.

$$\ddot{x} + x = \mu(1 - \beta x^2)\dot{x}$$

$$x(t) = ae^{it} + a^*e^{-it}$$

$$\dot{x}(t) = iae^{it} - ia^*e^{-it}$$

Substitution of the last two expressions into the equation for x leads to the equation for a:

$$\frac{da}{dt} = \frac{\mu}{2i}[1 - \beta(a^2 e^{i2t} + 2|a|^2 + a^{*2}e^{-i2t}](iae^{it} - ia^*e^{-it})e^{-it}$$

If $\mu = 0$, we have $da/dt = 0$, that is, the oscillations are harmonic. If $\mu \ll 1$, $a(t) \approx a(t+1)$. Let us integrate the equation over time using this fact.

$$\int_t^{t+1} \frac{da}{dt}\, dt = \Delta a = \frac{1}{2i}\int_t^{t+1}[a - \beta|a|^2\,a - \beta\,(a^3 e^{i2t} - a^{*3}e^{i4t} - 3|a|^2\,a^* e^{-i2t})]\, dt$$

Since $a(t)$ changes slowly, we can consider it as a constant under the integral, and in this case all terms with exponents will be zero after averaging.

Then the equation for $a(t)$ will be

$$\frac{da}{dt} = \frac{\mu}{2}[1 - 3\beta|a|^2]\,a.$$

or, if we introduce **slow** time $\tau = t\mu$,

$$\frac{da}{d\tau} = \frac{1}{2}[1 - 3\beta|a|^2]\,a$$

To obtain a real equation we substitute

$$a = Ae^{i\phi}$$

Then we can write the equation in gradient form:

$$\frac{dA}{d\tau} = -\frac{\partial U}{\partial A} \quad \text{with} \quad U = -\frac{A^2}{4} + \frac{3\beta A^4}{8}$$

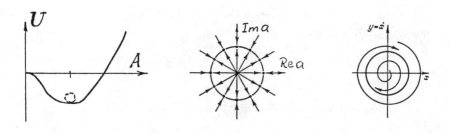

Figure 5.3

The state $A = 0$ is unstable, and as time evolves, the system will "roll down" to the state

$$A_0 = \sqrt{\frac{2}{3\beta}}$$

In the phase plane this will correspond to a limit cycle[5] with the radius A_0. We can easily write the solution for x:

$$x(t) = 2A_0 \cos t$$

To set any final worries at rest, let us now consider an example where we can find the exact solution. We start with a linear dissipative system:

$$\dot{x}_1 = \alpha x_1 - x_2$$

$$\dot{x}_2 = x_1 + \alpha x_2$$

The characteristic equation is

$$\begin{vmatrix} \alpha - \lambda & -1 \\ 1 & \alpha - \lambda \end{vmatrix} = (\alpha - \lambda)^2 + 1 = \alpha^2 - 2\alpha\lambda + \lambda^2 + 1 = 0.$$

$$\lambda_{1,2} = \alpha \pm \sqrt{-1} = \alpha \pm i.$$

If $\alpha < (>)\, 0$ we have a stable (unstable) focus. If $\alpha = 0$, we have a center. We are interested in the case when $\alpha > 0$ and oscillations grow. The disturbing feature of our system is that this growth is unbounded. But in Nature we do not observe infinite voltages, currents, fields, etc. Therefore, in the real world there is something else entering the right hand side of our equations: something that stops the growth of oscillations. Let us act as physicists and correct our proposed system. Introduce nonlinear terms which are proportional to $\beta(x_1^2 + x_2^2)$:

$$\dot{x}_1 = \alpha x_1 - x_2 - \beta\left(x_1^2 + x_2^2\right)x_1$$

[5]A stable limit cycle is an *attractor* in the sense that it draws into itself all phase trajectories from its neighborhood.

$$\dot{x}_2 = x_1 + \alpha x_2 - \beta \left(x_1^2 + x_2^2\right) x_2.$$

If we change to variables A and ϕ,

$$x_1 = A \cos \phi \quad x_2 = A \sin \phi.$$

we arrive at

$$\dot{x}_1 = \dot{A} \cos \phi - \dot{\phi} A \sin \phi = \alpha A \cos \phi - A \sin \phi - \beta A^3 \cos \phi$$

$$\dot{x}_2 = \dot{A} \sin \phi + \dot{\phi} A \cos \phi = a \cos \phi + \alpha A \sin \phi - \beta A^3 \sin \phi.$$

We can write the equations for A and ϕ:

$$\dot{A} = \alpha A - \beta A^3$$

$$\dot{\phi} A = A \Rightarrow \dot{\phi} = 1$$

We are left with the same equation as we derived approximately in the previous lecture, even though we did not make any approximations. This is a very important result.

Now let us discuss behavior in systems with nonlinear dissipation from an energy balance point of view. In Figure 5.4 we show energy supply and energy dissipation for some system. Quite obviously, if supply and dissipation are equal to each other the amplitude of oscillations cannot increase or decrease unless we firmly shake our system. Therefore, the positions of limit cycles are the intersections of those two lines.

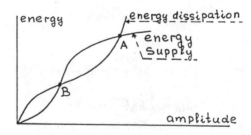

Figure 5.4

Look at point A. What will happen if we do shake the system? Assume A changed to $A + \delta A$.[6] We can see that the energy leak is greater than the energy supply, and, therefore, the amplitude will decrease until it reaches A. In point B, on the contrary, the energy supply becomes greater than the losses as the system moves from B to $B + \delta B$ ($\delta B > 0$), and the amplitude keeps growing. Thus A corresponds to a stable limit cycle and B, to an unstable one.

---

[6]We assume that $\delta A > 0$, but obviously the qualitative discussion will be the same if $\delta A < 0$.

6 The Poincaré Map

In the previous lectures we saw that sometimes we can reduce a two-dimensional problem to a one-dimensional one by using the fact that the motion can be divided into a very fast part and a much slower part. Maybe it does not look that important now because we just made a simple problem simpler. But think a little bit about how you would describe a system with three-dimensional phase space. This is already a very difficult problem. And can you imagine four- or five-dimensional phase spaces? How wonderful (and useful!) it might be if we could manage to find some trick to reduce the dimensions of these spaces at least by one. As a matter of fact, sometimes we really can do this! To understand how we can do this let us back up to limit cycles.

We noticed already that when we wrote the equation for an amplitude near a limit cycle, we intentionally got rid of extra information about the phase. One way to treat this is to say that the equation describes motions not in whole plane but just on the line 0Σ (Figure 6.1). The trajectory crosses 0Σ at discrete moments of time. 0Σ is referred to as the secant line. Therefore, if we have a set of initial conditions for amplitudes on 0Σ, namely a whole interval on 0Σ, the phase flow in the plane will perform a map of this set over the period as shown in Figure 6.1. This map is from the line crossing the full flow back to that same line. This map is named after Henri Poincaré who first introduced this method in his studies of the three body problem in celestial mechanics.

In general, a dynamical system with discrete time evolution may be described by the following equation[7]:

$$\mathbf{x}(k) = \mathbf{F}(\mathbf{x}(k-1))$$

---

[7]Note that the definition of dynamical system does not depend on whether it is continuous in time or discrete. In our discrete system the value of the dynamical variable at time k is uniquely defined by its value at the previous moment of time $k-1$.

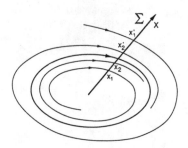

Figure 6.1

Let us consider the system we studied in the previous lecture.

$$\frac{dA}{d\tau} = -[\gamma - \alpha' A^2 + \beta' A^4]A$$

τ is the slow time, and A changes very little over the period. Therefore, we can approximate the drivative with respect to τ as

$$\frac{dA}{d\tau} \approx \frac{A(\tau) - A(\tau - 1)}{\Delta\tau}$$

And finally we can write the equation for A:

$$A(\tau) = (1 - \gamma)\,A(\tau - 1) + \alpha'\,A^3(\tau - 1) - \beta' A^5(\tau - 1)$$

There is a very simple method to qualitatively describe motions in a one-dimensional map. Look at Figure 6.2:

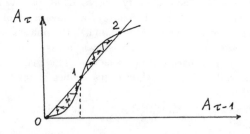

Figure 6.2 $(0 < \gamma < 1)$

Assume you have some initial point on the horizontal axis. Obviously, to find the value at the next moment of time you just need to draw a vertical line to the point where it intersects the function curve $A(\tau) = F(A(\tau - 1))$. Then, if you draw a horizontal line to the point where it intersects the bisector, the horizontal coordinate of this point will give you the initial value for the next step.[8]

We can easily find the fixed points of the map. These are the points where A at time τ is the same as A at time $\tau - 1$, that is the point where the function curve crosses the bisector. If you choose the initial condition somewhere near point 1 the system will go away from that point. This means that point 1 is unstable. In the same way you can find that point 2 is a stable one. One can establish that a fixed point is stable if at the fixed point we have

$$\left| \frac{dA(\tau)}{dA(\tau - 1)} \right| < 1$$

and is unstable if the opposite condition is fulfilled.

Now let's go for real excitement. Let us look at what is going on near a limit cycle in three-dimensional phase space. In this case we should consider the points where trajectory crosses a secant surface, not just a secant line as we did it for two-dimensional case. The choice of secant surface in general is not unique, but for our case there is one very convenient choice, namely, a plane. Let us assume that the limit cycle lies on a cylinder.

Then the best way is to introduce the secant plane as the plane passing through the axis of the cylinder as we show in Figure 6.3.

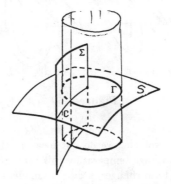

Figure 6.3

This Poincaré section really reduces the problem by one dimension. On the Poincaré plane the evolution is discrete in time, but we still can see all known types of equilibrium points (Figures 6.4 and 6.5): node, focus, saddle, center.

One may think of systems with discrete time as just a good way to study continuous systems. But this is not quite correct. In real life there are a lot of things which we do not measure continuously. Biologists measure the fly population on Earth not

---

[8]This procedure is sometimes called Lamere stairs (ladder).

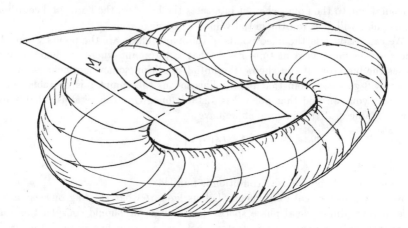

Figure 6.4

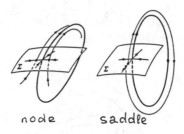

node   saddle

Figure 6.5

every hour and even not every day. They make measurements once a year. If we study the global warming problem, we do not measure the temperature every five seconds. We measure the average temperature this year, next year and so forth. For everyone who studies long term changes, a discrete model is not an auxiliary method but an independent subject of her research.

Before we finish for today let us consider one very important map, namely the so-called logistic map. It was introduced to describe changes in species population and later became an important example to illustrate that even in this quite simple one-dimensional system one can observe very complicated behavior. Figures 6.6, and 6.7 show how the system

$$x(n+1) = \lambda x(n)(1 - x(n))$$

behaves at different values of λ.

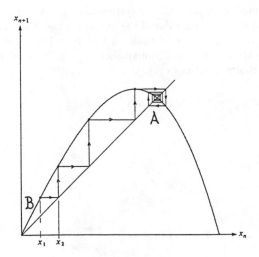

Figure 6.6

And finally:

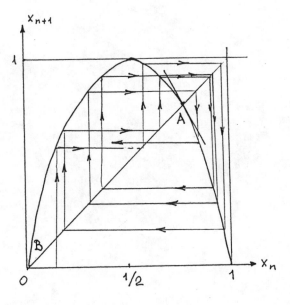

Figure 6.7

If the control parameter is small, the system has one stable (A) and one unstable (B) fixed point. At some critical value of the parameter the slope at point A is exactly -1, and the point is not stable or unstable. If the control parameter is large enough both fixed points are unstable and the system bounces between them in a chaotic manner: it never comes to a limit cycle or a fixed point!

7 Slow and Fast Motions in Systems with One Degree of Freedom

Let us recall of one of the previous lectures when we considered a linear oscillator with very large dissipation (or a pendulum in a very viscous liquid): a system which is very close to a gradient one. The equation for such a system is

$$\mu\ddot{x} + \dot{x} + x = 0$$

or

$$\begin{aligned} \dot{x} &= y \\ \mu\dot{y} &= -y - x \end{aligned} \qquad \text{where } \mu \ll 1$$

We know that we can consider two kinds of motion in this system: fast motions shown by the vertical lines on the phase plane, and slow motions shown by the line $x = -y$ (see Figure 7.1).

We can write separate equations for these two motions. For fast motions, the equation will be $x \approx$ const, and for slow motions $dx/dt = -x$. This separation of two motions obviously simplifies the problem effectively. The system passes through the region of fast motions very quickly, and after that the problem becomes one-dimensional.

Now let us think what will happen if we have a nonlinear system. The classical example which illustrates fast and slow motions in a nonlinear system is the Van der Pol equation with very strong dissipation:

$$\ddot{z} - \epsilon(1 - \beta z^2)\dot{z} + z = 0 \quad \epsilon \gg 1$$

Unfortunately, this equation is not too convenient to look at. Let us change to the variable v defined by $z = dv/dt$

$$\frac{d^3 v}{dt^3} - \epsilon(1 - \beta\dot{v}^2)\ddot{v} + \dot{v} = 0$$

33

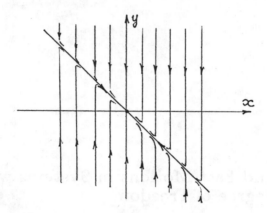

Figure 7.1

One can integrate this system:

$$\frac{d}{dt}\left[\ddot{v} - \epsilon\left(1 - \frac{\beta}{3}\dot{v}^2\right)\dot{v} + v\right] = 0$$

$$\ddot{v} - \epsilon\left(1 - \frac{\beta}{3}\dot{v}^2\right)\dot{v} + v = \text{constant}$$

We can always get rid of the constant on the right hand side by $v \to v+$ constant. We will assume for the sake of simplicity that $\beta = 3$.

$$\ddot{v} - \epsilon(1 - \dot{v}^2)\dot{v} + v = 0$$

This is the famous Rayleigh equation. Now we will make our final change of variables: $t' = t/\epsilon$, $x = v/\epsilon$. This leads us to

$$\epsilon^{-2}\ddot{x} - (1 - \dot{x}^2)\dot{x} + x = 0$$

or

$$\mu\ddot{x} - (1 - \dot{x}^2)\dot{x} + x = 0 \text{ where } \mu = \epsilon^{-2} \ll 1$$

or in another form

$$\dot{x} = y$$
$$\mu\dot{y} = (1 - y^2)y - x$$

It is easy to write the equation for the integral curves:

$$\frac{dy}{dx} = \frac{(1 - y^2)y - x}{\mu y}$$

Since $\mu \ll 1$, dy/dt is very large unless $(1 - y^2)y - x = 0$. Therefore, everywhere except on the line $x = (1 - y^2)y$ the trajectories will be vertical lines describing fast motions. The direction of motion at some point is defined by the sign in the inequality

$$x_0 > (1 - y_0^2)y_0 \text{ or } x_0 < (1 - y_0^2)y_0$$

because $\mu dy/dt = (1 - y^2)y - x$.

The system moves along the fast motion lines and eventually comes into the region where $x \approx (1 - y^2)y$. This is the region of slow motions. The equation $dx/dt = y$ means that on this line x decreases when $y < 0$ and increases when $y > 0$.

Now we are in position to draw the phase portrait of the system.

Let us follow a point on some trajectory (point E on Figure 7.2). First it moves very fast from E to C and then it slowly crawls to D. Note that along CD the slow motion is stable with respect to fast motions. If you kick a point away from the line of slow motions, it will come back very quickly. As the system reaches point D the situation changes. What will happen if you push the system in the positive y direction? It will leave the slow motion region and will be carried away by the fast motion line! This line will bring the system to point A and everything will start again. It is easy to see that no matter where the point starts from it will eventually find itself on the limit cycle ABCD which is therefore stable. Figure 7.3 shows the function $y(t)$ when the system oscillates in the limit cycle configuration.[9]

You can see that the system spends almost all the time on the line of slow motions. This fact simplifies the problem enormously because the motion on this line can be described by just one first order equation:

$$\dot{x} = y; \ x \approx (1 - y^2)y$$

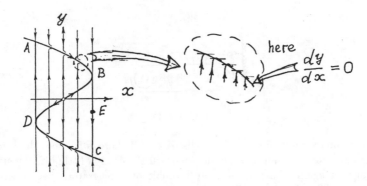

Figure 7.2

---

[9]Oscillations of this kind are often called relaxation oscillations.

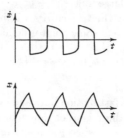

Figure 7.3

$$dt = \frac{dx}{y} \approx \frac{1}{y} d\left[(1 - y^2)y\right]$$

$$y \approx \dot{y} - 3y^2\dot{y} \Rightarrow \dot{y} \approx \frac{y}{1 - 3y^2}$$

It is clear that we can write for the period of oscillations

$$T = T_{AB} + T_{CD}$$

Furthermore, since our system is symmetric with respect to $y \to -y$,

$$T_{AB} = T_{CD} \text{ and } T = 2T_{AB}$$

$$T = 2 \int_{y_A}^{y_B} \frac{d(y - y^3)}{y} = 2\left(\ln y - \frac{3}{2}y^2\right)\Big|_{y_A}^{y_B}$$

Now let us consider one more important example of a system where one can observe fast and slow motions: a brick on a conveyor belt.

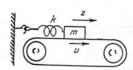

Figure 7.4

Our common sense and experience tells us that if there is friction, the brick will move with the conveyor for a while but then will slip back a little bit. Then it will repeat all over again. This is one more example of a generator. Let us write the equations of motion for the brick.

$m\ddot{z} = -kz + \mid! F_{\text{friction}}(u)$ where $u = \dot{z}$ and the function F_{friction} is shown in the Figure

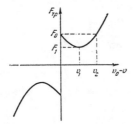

Figure 7.5

After the change of variables

$$x = \frac{kv^2}{F_0} z; \; y = \frac{u}{v}; \; \tau = \frac{kv}{F_0} t; \; f = \frac{F_{\text{friction}}}{F_0}; \; \mu = \frac{mkv^2}{F_0^2},$$

we arrive at the system

$$\mu \frac{dy}{d\tau} = -x - f(y - 1)$$

$$\frac{dx}{d\tau} = y$$

If $\mu \ll 1$ we will see completely the same behavior as we saw in the Van der Pol generator. We think that using the knowledge from this lecture you can analyze the system yourself. (The answer is in Figure 7.6.)

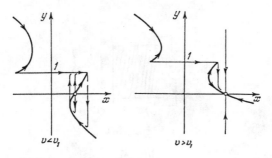

Figure 7.6

8   Forced Nonlinear Oscillators:
Linear and Nonlinear Resonances

So far we have only considered systems which did not depend on time explicitly: these systems are called autonomous. Now let us see what will happen when a nonlinear system is subject to an external force. An example of such a system is shown in Figure 8.1 (an oscillating ferromagnetic ball in a time dependent magnetic field).

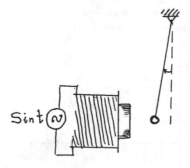

Figure 8.1

We will consider only relatively small oscillations in a system with weak dissipation and weak external force. In this case we can expand the nonlinearity in the equations of motion for the displacement of the pendulum $x(t)$ in a power series and keep just the three leading terms:

$$\ddot{x} + \omega_0^2 x = \mu[E \cos\omega t - \alpha x^2 - \beta^3 - 2\gamma\dot{x}]; \quad \mu \ll 1$$

If $\alpha = \beta = 0$, we have the usual forced, damped linear oscillator. Let us recall briefly what we will see in this case. Since the system is linear, we do not need to assume $\mu << 1$ to solve the problem

$$\ddot{x} + 2\gamma\dot{x} + \omega_0^2 x = E \cos\omega t$$

Let us first consider the resonance in the undamped oscillator $(\gamma = 0)$. We can write the general solution in the form

$$x(t) = A \cos\omega_0 t + B \sin\omega_0 t + \frac{E}{\omega_0^2 - \omega^2}\cos\omega t$$

For example, for the initial conditions $x = dx/dt = 0$ at $t = 0$, we have the solution

$$x(t) = \frac{E}{\omega_0^2 - \omega^2}\left(\cos\omega t - \cos\omega_0 t\right)$$

This is a quasi-periodic motion with frequencies

$$\Omega_{1,2} = \frac{\omega_0 \pm \omega}{2}$$

whose behavior is shown in Figure 8.2

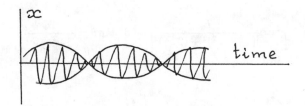

Figure 8.2

However, if the external frequency is close to the natural frequency of the system $(\omega \approx \omega_0)$ we must be more careful. In this case we set

$$\omega \to \omega_0 \Rightarrow \omega_0^2 - \omega^2 = (\omega_0 - \omega)(\omega_0 + \omega) \approx 2\omega_0(\omega_0 - \omega)$$

and

$$x(t) = \frac{E}{2\omega_0(\omega_0 - \omega)}2\sin\omega_0 t \sin\frac{(\omega_0 - \omega)t}{2} =$$

$$\frac{Et}{2\omega_0}\sin\omega_0 t \frac{\sin\frac{\omega_0 - \omega}{2}t}{\frac{\omega_0 - \omega}{2}t} \to \frac{Et}{2\omega_0}\sin\omega_0 t$$

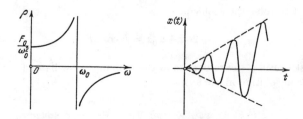

Figure 8.3

The system is unstable. But we do not see an exponential instability, as we saw before. In this system the growth is linear with time. This called secular growth.

What will change if we have dissipation? The equation will be

$$x = ReX; \quad \ddot{X} + 2\gamma\dot{X} + \omega_0^2 X = Ee^{i\omega t}$$

If we wait long enough the dissipation will kill the natural oscillations, and the only motion we will observe will be due to the forcing. We can write this solution in the form

$$X = \rho e^{i(\omega t - \theta)},$$

and find

$$\rho^2 = \frac{E^2}{[(\omega_0^2 - \omega^2)^2 + (2\omega\gamma)^2]} \quad \text{and} \quad \tan\theta = -\frac{2\omega\gamma}{\omega_0^2 - \omega^2}$$

This is shown in Figure 8.4:

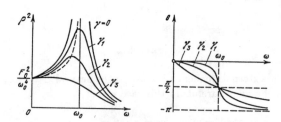

Figure 8.4

Now, what do we expect to see in the case of nonlinear resonance? First of all, if we fix the external frequency we will not see the amplitude grow to infinity even in the case when we have no dissipation. It is clear why. Assume the frequency were exactly equal to that of linear oscillations. As the nonlinear oscillations grow the dependence of the period of oscillations on their amplitude starts playing an important role. The amplitude of the external force no longer matches the intrinsic frequency; the system goes out of resonance and the oscillations stop growing. From this point of view the words "exact resonance" do not have any physical sense for the nonlinear system.

Assume the nonlinear system is conservative, and the frequency of free oscillations increases with amplitude. Then it is clear that the larger the external frequency, the larger the amplitude of forced oscillations in the system can be.

Another important feature of nonlinear resonance is that we will see many resonance peaks (see Figure 8.5 for example).

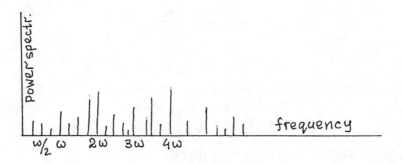

Figure 8.5

This happens because in a nonlinear system the undamped solution in not just sine or cosine but a Fourier series. Therefore we would expect to see resonance peaks at frequencies $2\pi n/T$ where T is the period of undamped oscillations. A very convenient way to analyze nonlinear resonance in a system with small nonlinearity, small dissipation and small external force is to consider the average work done by the external force and dissipative force. This work is approximately equal $\int_t^{t+T}[E\cos\omega t - \gamma\dot{x}]\dot{x}dt$. When the average work is positive the energy is pumped into the system and the oscillation grows.

Let us do some quantitative analysis for μ small and for the external frequency close to that of the linear oscillations. We will limit ourselves to just the first (main) resonance. In this case we can write

$$\omega_0^2 = (\omega - \mu\epsilon)^2 = \omega^2 - 2\mu\epsilon\omega + \mu^2\epsilon^2 \approx \omega^2 - 2\mu\epsilon\omega \text{ because } \mu \ll 1$$

and the equation of motion will be

$$\ddot{x} + \omega^2 x = \mu[E\cos\omega t - \alpha x^2 - \beta^3 - 2\gamma x + 2\epsilon\omega x].$$

We can use the method we have developed before to solve an equation written in the form

$$\ddot{x} + \omega^2 x = \mu f(x, \dot{x}, t) \quad \mu \ll 1.$$

We are looking for the solution in the form

$$x = a(t)e^{i\omega t} + a^*(t)e^{-i\omega t}.$$

If we write $a(\mu t) = A(\mu t) \cdot \exp(i\Phi(\mu t))$, we find the approximate equations for amplitude and phase:

$$\frac{dA}{d\tau} = \frac{E}{2}\omega \cos \Phi - \gamma A$$

$$\frac{d\Phi}{d\tau} = -\frac{E}{2\omega A} \sin \Phi + 3\frac{\beta}{8}\omega A^2 - \epsilon \quad \text{where} \quad \tau = \mu t$$

The equilibrium states of this system are defined by

$$\gamma A_0 = \frac{E}{2\omega} \cos \Phi_0; \quad -\epsilon + \frac{3\beta A_0}{8\omega} = \frac{E}{2\omega A_0} \sin \Phi_0$$

$$\text{or} \quad \gamma^2 + (\delta A_0^2 - \epsilon)^2 = \frac{E^2}{4\omega^2 A_0} \quad \text{where} \quad \delta = \frac{3\beta}{8\omega}.$$

And at last the equation for the resonance curve will be

$$\epsilon = [\delta A_0]^2 \pm \sqrt{\frac{E^2}{4A_0^2\omega^2} - \gamma^2}.$$

Figure 8.6 shows this dependence for the conservative case.

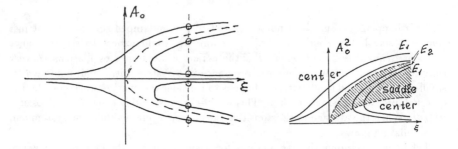

Figure 8.6

The shape of this curve confirms our qualitative speculations. But we see something more to it. Unlike the linear resonance the nonlinear resonance demonstrates three equilibrium states for A_0^2 (or six for A_0) for each value of frequency detuning! This is a very interesting and important phenomenon. Which of these three states will the system choose to stay in? If you do the standard linear instability analysis, you will find out that in the shadowed region we have a saddle-type equilibrium state. The two other states are centers. Figure 8.7 shows the phase space for this system in coordinates $Re(a)$ and $Im(a)$. All trajectories inside the separatrix correspond to the case when the oscillations are synchronized with the external force. And the trajectories outside the separatrix correspond to motions out of resonance. Note that this diagram has almost nothing in common with the corresponding diagram for a linear oscillator.

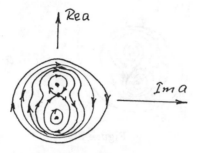

Figure 8.7

Now let us look at Figure 8.8 which shows the resonance curve for a dissipative oscillator.

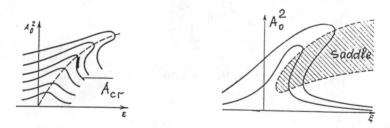

Figure 8.8

We see that if the dissipation is large enough, there is just one equilibrium amplitude for any frequency. The larger the dissipation the closer the curve to that of a linear oscillator.

We can easily construct the phase plane $(\mathrm{Re}(a), \mathrm{Im}(a))$, if we recall that dissipation turns centers into foci and leaves saddles unchanged (Figure 8.9).

There is one more interesting question, namely, "What will happen if we change the frequency detuning very slowly on the time scale of all other changes?" Well, as long as the system can stay in the stable focus A it will stay there. But at some moment the focus "annihilates" with the saddle, and the system jumps to another focus. If we now decrease the detuning $(\omega - \omega_0)$, the system will stay in that new "house" until it is destroyed by the collision with the saddle point. At this moment the amplitude of oscillations will suddenly increase. These two jumps form the hysteresis loop on the resonance diagram.

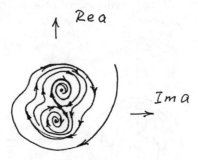

Figure 8.9

This hysteresis loop is shown in Figure 8.10:

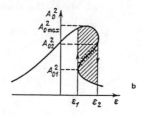

Figure 8.10

9 Forced Generator: Synchronization

In the previous lecture we became acquainted with a few very interesting phenomena taking place in a forced nonlinear oscillator: nonlinear resonance (periodic oscillations with the frequency equal to that of the external force), pulsation (quasi-periodic oscillations with two incommensurate frequencies whose phase space portrait consists of a dense winding on a two-dimensional torus), multistability (more than one stable stationary solutions corresponding to the resonance).

What shall we see if we apply a periodic external force to a generator? In this lecture we will deal with the simplest case when the generator has just one limit cycle, and, therefore, the parameters of the asymptotic behavior of the system do not depend on the initial condition.

The main phenomenon in this system is called frequency locking; that is, onset of oscillations with frequency equal to that of the external force. Probably, the first scientist who paid attention to this phenomenon was Huygens! He discovered the following trick to help an owner of a clock shop get rid of a bad clock. He noticed that if you have a very big and heavy precise clock attached to the same wall that a smaller and less precise clock is hanging on, the second clock will oscillate at the frequency of the first one. As soon as you sell the small clock to somebody and take it from that wall it will no longer show the exact time. The trick is that the big clock shakes the wall a little bit and the second clock "feels" this and somehow changes its frequency. Now let us see how all this happens.

The simplest model one can think of is the forced Van der Pol generator

$$\ddot{x} - \mu(1 - x^2)\dot{x} + (1 + \mu\xi)x = \mu E' \cos t, \tag{13}$$

or in first order form

$$\begin{aligned}
\dot{x} &= y \\
\dot{y} &= = -x + \mu f(x, y, t), \tag{14}
\end{aligned}$$

45

where

$$f(x, y, t) = E' \cos t - \xi x + (1 - x^2)y. \tag{15}$$

Now $\mu \ll 1$, and the frequency detuning $\mu\xi \ll 1$.

Here we assumed for simplicity the frequency of the external force to be unity. The approximate expression for the eigenfrequency of the generator is the same as in the previous lecture.

After we make our standard substitution

$$\begin{aligned} x(t) &= a(t)e^{it} + a^*(t)e^{-it} \\ y(t) &= i[a(t)e^{it} - a^*(t)e^{-it}], \end{aligned} \tag{16}$$

and average the equation for a(t), we can write the following equations for the slow variables

$$A = \frac{\mathrm{Re}a}{2} \ ; \ B = \frac{\mathrm{Im}a}{2}. \tag{17}$$

Then we arrive at $(E' = 2E)$

$$\begin{aligned} \dot{A} &= A[1 - (A^2 + B^2)] - B\xi + E \\ \dot{B} &= B[1 - (A^2 + B^2)] + A\xi. \end{aligned} \tag{18}$$

If E = 0 we have the standard equations for a Van del Pol generator, except for those strange ξ's on the right. Why do we have them? The answer is that we chose the wrong frequency for the averaging method. Instead of $1 + \mu\xi$ we chose just 1. This is not a big deal, because if we multiply the slow function A(t) by $e^{i\mu\xi t}$, the result will be still a slow function as long as $\mu \ll 1$. Therefore the averaging method still works and ξ is just compensation for such a strange choice of the frequency. On the phase plane (A,B) the terms with ξ correspond to rotation with the frequency ξ (see Figure 9.1). The phase portrait in (x,y)-coordinates, of course, will be the same in both cases

Now let us go back to the case when E is nonzero. As usual we start with an analysis of fixed points. In this case the fixed points correspond to complete synchroni-

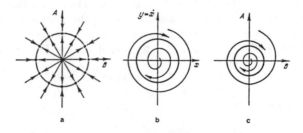

Figure 9.1

zation (the small clock oscillates at the frequency of its big sister). The equilibrium states are determined by

$$A_0(1 - (A_0^2 + B_0^2)) - B_0\xi = = -E$$
$$B_0(1 - (A_0^2 + B_0^2)) + A_0\xi = 0. \tag{19}$$

After we square these equations and add them to each other we find the equation for the intensity of the oscillations $\rho = A^2 + B^2$:

$$\rho(1 - \rho)^2 + \xi^2\rho = E^2. \tag{20}$$

We can solve this equation for the detuning ξ:

$$\xi = \sqrt{\frac{E^2 - \rho(1 - \rho)^2}{\rho}}. \tag{21}$$

ξ as a function of ρ for different values of E is shown in Figure 9.2. Obviously, real values of ξ exist only if $E^2 \geq \rho(1 - \rho)^2$. We can see that if $E^2 < \frac{4}{27}$, (the external force is weak) there are two separate branches of $\xi(\rho)$. If the external force is strong, there is just one branch.

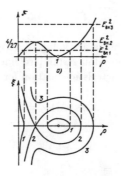

Figure 9.2

The example of nonlinear resonance shows that some of these branches can be unstable, and only those which are stable correspond to a synchronization regime. Therefore the next thing to do is to study the linear instability of these stationary points.

Before we write any equations let us think what we already know about the stability. First of all, we know how the trajectories at infinity behave. For these trajectories we can neglect terms with E, $A\xi$, $B\xi$, A and B because they are small (at large B and A) compared with terms A^3, B^3, AB^2 and BA^2. After we do this we will be left with the evolution equations

$$\dot{A} = = -A(A^2 + B^2)$$
$$\dot{B} = -B(A^2 + B^2).$$
(22)

This leads to

$$\frac{dA}{dB} = \frac{A}{B}.$$
(23)

From these two equations it is clear that A and B will decrease and that the phase portrait at large amplitudes will look like a degenerate node or focus (Figure 9.3).

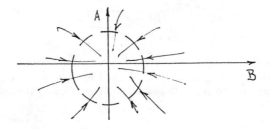

Figure 9.3

But the most interesting things are going to happen in the region close to zero. In this region we can not escape doing standard linear instability analysis. For the problem we have here the characteristic equation for the stability exponents is

$$\lambda^2 + p\lambda + q = 0,$$
(24)

where

$$p = 4\rho - 2$$
$$q = (1 - 3\rho)(1 - \rho) + \xi^2.$$
(25)

Now we can partition the (ξ, ρ) plane into regions with different types of stability (Figure 9.4).

The states above the thick line are stable, below, unstable. The stable equilibrium states correspond to the synchronization regime as we discussed before.

Now we have an interesting opportunity to construct the entire phase plane (A,B) for a given external amplitude and the detuning. To do so we must superimpose the diagram for the intensities of equilibrium states for different external amplitudes and detuning with the diagram of their stability (Figure 9.5).

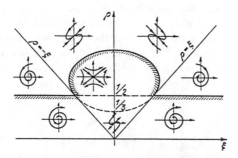

Figure 9.4

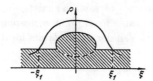

Figure 9.5

Let us first consider the case of strong external signal $E^2 > \frac{8}{27}$. In this case we have just one equilibrium state at any detuning.

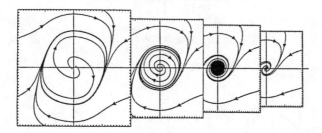

Figure 9.6

This diagram also indicates that this equilibrium state is either a node or a focus. Let us discuss what will happen if we increase the detuning a little bit beyond the stability region. Obviously the stable focus turns into an unstable one, and we do not have any stable fixed points. But we know that trajectories on infinity are still coming into the region close to zero. The only way how we can resolve the problem of incoming trajectories is to draw a limit cycle near the origin. This is the famous Andronov-Hopf bifurcation when a stable focus gives birth to an unstable focus and a stable limit cycle. See Figure 9.6. Then the synchronization ceases to exist, and a regime of pulsations appears. This is the "soft" excitation for the amplitude of modulation and a hard excitation for the frequency. These are shown in Figure 9.7:

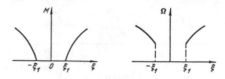

Figure 9.7

What will happen in the case of weak signals? Figure 9.8 shows the detuning range for which there exist a stable fixed point corresponding to the synchronization regime. The phase portraits in Van der Pol variables for different detunings are shown in Figure 9.9

The detuning range for synchronization:

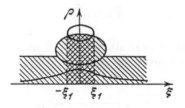

Figure 9.8

The phase portraits for various ξ:

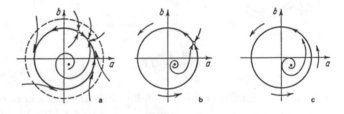

Figure 9.9

10 Competition of Modes

Today we are going to discuss how two coupled generators behave. This problem is very important. This problem appears in many branches of physics (coupled modes in a laser resonator, in convective flow in a fluid, in electrical circuits and so on).

In a simple formulation competition is a purely energetic phenomenon. It does not involve the phase of oscillations. Furthermore, it occurs even in some non-oscillatory systems (for example, competition between biological species).

This phenomenon is pretty complicated and, therefore, it is logical to study the simplest system that is a two-frequency Van der Pol generator (Figure 10.1).

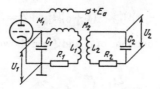

Figure 10.1

From the Kirchhoff circuit rules, we can find the equations for this system

$$L_1 C_1 \ddot{U}_1 + [R_1 C_1 \; - \; MS(U_1)]\dot{U}_1 + U_1 + NC_2 \ddot{U}_2 = 0$$
$$L_2 C_2 \ddot{U}_2 + R_2 C_2 \dot{U}_2 \; + \; U_2 + NC_1 \ddot{U}_1 = 0, \tag{26}$$

where

$$S(x) = S_0 - S_2 x^2. \tag{27}$$

After we change variables to

$$t' = \frac{1}{\sqrt{L_1 C_1}} t \; ; \quad X_1 = U_1 \sqrt{\frac{MS_2}{MS_0 - R_1 C_1}} \; ; \quad X_2 = U_2 \frac{L_2 C_2}{NC_1} \sqrt{\frac{MS_2}{MS_0 - R_1 C_1}}, \tag{28}$$

51

and introduce new parameters

$$n_{1,2}^2 = \frac{1}{L_{1,2}C_{1,2}} \ ; \ \xi = \frac{n_2^2}{n_1^2} \ ; \ \alpha = \frac{N^2}{L_1 L_2} < 1, \tag{29}$$

and

$$\mu = n_1(M S_0 - R_1 C_1) \ ; \ \delta = \frac{R_2 L_1 C_1}{L_2(M S_0 - R_1 C_1)}, \tag{30}$$

we can write these equations in a simpler form:

$$\begin{aligned} \ddot{X}_1 + X_1 + \alpha \ddot{X}_2 &= \mu(1 - X_1^2)\dot{X}_1 \\ \ddot{X}_2 + \xi X_2 + \ddot{X}_1 &= \mu\delta\dot{X}_2. \end{aligned} \tag{31}$$

If $\mu = 0$, we have a linear problem. We look for the solution in the form

$$\begin{aligned} X_1 &= a_1 e^{i\omega_1 t} + a_2 e^{i\omega_2 t} + cc \\ X_2 &= a_1 \Psi_1 e^{i\omega_1 t} + a_2 \Psi_2 e^{i\omega_2 t} + cc, \end{aligned} \tag{32}$$

then we find that

$$\Psi_{1,2} = \frac{1 - \Omega_{1,2}^2}{\alpha \Omega_{1,2}^2} = \frac{\Omega_{1,2}^2}{\xi - \Omega_{1,2}^2}, \tag{33}$$

where

$$\Omega_{1,2} = \frac{\omega_{1,2}}{n_1}. \tag{34}$$

The Ω are, therefore, the roots of

$$(1 - \alpha)\Omega_{1,2}^4 - (1 + \xi)\Omega_{1,2}^2 + \xi = 0. \tag{35}$$

The eigenfrequencies as functions of α and ξ are shown in Figure 10.2

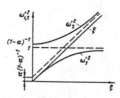

Figure 10.2

Now let us think what we would expect to see when μ is not zero but very small. Our experience says that we may try to look for the solution in the form

$$\begin{aligned} X_1 &= a_1(\mu t)e^{i\omega_1 t} + a_2(\mu t)e^{i\omega_2 t} + cc \\ X_2 &= a_1(\mu t)\Psi_1 e^{i\omega_1 t} + a_2(\mu t)\Psi_2 e^{i\omega_2 t} + cc \end{aligned} \tag{36}$$

When μ is small the amplitudes are slow functions of time, and we can employ the averaging method to obtain the following equations:

$$\frac{da_1}{dt} = \frac{\mu}{2}h_1[1 - (|a_1|^2 + \rho_{12}|a_2|^2)]a_1$$

$$\frac{da_2}{dt} = \frac{\mu}{2}h_2[1 - (|a_1|^2 + \rho_{21}|a_2|^2)]a_2, \tag{37}$$

with coefficients

$$\sigma_1 = \omega_1^4 \frac{\omega_2^2 - 1}{4(\omega_2^2 - \omega_1^2)} \tag{38}$$

$$\sigma_2 = \omega_2^4 \frac{1 - \omega_1^2}{4(\omega_2^2 - \omega_1^2)} \tag{39}$$

$$\lambda_1 = 4(1 - \frac{\delta\Psi_1}{\xi\Psi_2}) \tag{40}$$

$$\lambda_2 = 4(1 - \frac{\delta\Psi_2}{\xi\Psi_1}) \tag{41}$$

$$h_1 = \sigma_1\lambda_1 \; ; h_2 = \sigma_2\lambda_2 \tag{42}$$

$$\rho_{12} = 2\frac{\lambda_2}{\lambda_1} \; ; \rho_{21} = 2\frac{\lambda_1}{\lambda_2}. \tag{43}$$

It is easy to write the equations for the intensities of the oscillations $m_{1,2} = |a_{1,2}|^2$:

$$\dot{m}_1 = h_1[1 - m_1 - \rho_{12}m_2]m_1$$

$$\dot{m}_2 = h_2[1 - m_2 - \rho_{21}m_1]m_2. \tag{44}$$

Here we introduced the new time μt. These are often known as the kinetic equations.

Now at last we are done with the trivial, but boring, mathematics and can turn back to physics. Why is the phenomenon of mode competition so important? In our simple example we have just two oscillators and maybe the problem does not look too hard. But what shall we see in a system with many degrees of freedom? If all degrees of freedom are excited we will probably see very complicated behavior with very many lines in the power spectrum. But there is another possibility, namely, due to mode competition one mode suppresses all others and the system behaves in a relatively simple way, there are just a few spectral lines and so on. Let us illustrate how this may happen in our system. For simplicity let us assume that the coupling coefficients $\rho_{12} = \rho_{21}$ and also that each is equal to 2. Also take $h_1 = h_2$.

Then the equilibrium points are

$$m_1 = m_2 = 0 \; ; \; m_1 = 0, \, m_2 = 0 \; ; \; m_2 = 0, \, m_1 = 1 \; ; \; m_1 = m_2 = \frac{1}{3}. \tag{45}$$

It is also easy to see that the line $m_1 = m_2$ is an integral line (just divide one m-equation by another).

The phase plane of this system is shown in Figure 10.3.

Figure 10.3

What can we say about the behavior of the system? The system has just two stable equilibrium states. Therefore from any initial conditions after a possible transient, the system will come to one of them and will stay there forever. It means that by the end of transients only one mode will be excited. If we start with two modes whose amplitudes are different, the one with larger amplitude survives and suppresses another.

If we had two modes with different frequencies the picture would not change much. For some initial conditions first frequency wins, for another the second does. This is what we call mode competition.

The phase plane in the general case when $\rho_{12} \neq \rho_{21}$ is shown in Figure 10.4. At small values of these coupling coefficients the modes do not much affect each other, and they evolve independently. When $\rho_{12} > 1$ and $\rho_{21} < 1$, we have only one stable region. In this circumstance the action of one mode on the other may be strong while the inverse effect may be weak.

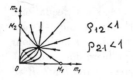
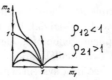

Figure 10.4

11   Poincaré Indices and Bifurcations of Equilibrium States

In this lecture we will consider a qualitative method which may help us escape sometimes very boring mathematics. It often leads us directly to an answer about the stability of equilibrium points, limit cycles and is very handy as we will see.

For a general system with two degrees of freedom we can write the following equations:

$$\begin{aligned}
\dot{X} &= P(X,Y) \\
\dot{Y} &= Q(X,Y).
\end{aligned} \tag{46}$$

Let us consider a closed curve N in the phase plane which does not pass through any equilibrium state (Figure 11.1).

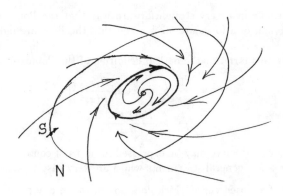

Figure 11.1

At any point S on this curve we can draw the vector $(P(x,y),Q(x,y))$ which is tangent to the phase trajectory at that point. If we move the point around the curve, the vector will rotate. By the time when the point returns after one rotation into its original position, the vector have experienced a certain (integer) number j of revolutions. We will define j to be positive, if the direction of revolution of the vector coincides with the direction of rotation of the point around N. To be more specific, we will assume that the point moves around the curve in a counterclockwise direction. Thus j can be zero or ±an integer.

What will happen if we change the curve a little bit? If N changes continuously and does not capture or shed any equilibrium points the angle of rotation will also change continuously. But the angle can be only $2\pi j$ where j is integer. Thus the angle and therefore j can not change. j does not depend on the shape of the curve. Therefore, all closed curves surrounding the same set of singular points will be characterized by the same j. This number j is called the Poincaré index.

For example, let us consider simple equilibrium points: center, node, saddle and focus (Figure 11.2).

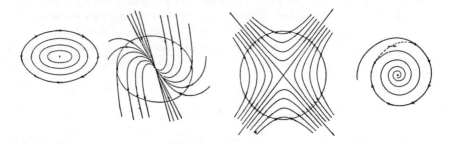

Figure 11.2

We can determine in a very straightforward way that the Poincaré indices of a center, of a node and of a focus are all +1, and that the Poincaré index of a saddle is -1.

We can also give a strict mathematical definition of the Poincaré indices:

$$
\begin{aligned}
j &= \frac{1}{2\pi} \oint_N d\arctan[\frac{Q(x,y)}{P(x,y)}] \\
&= \frac{1}{2\pi} \oint_N \frac{PdQ - QdP}{Q^2 + P^2}
\end{aligned}
\tag{47}
$$

So the index of a closed curve may be expressed in terms of contour integral.

We can make a few general statements which are very easy to verify:

1. The index of a closed curve which does not surround any fixed points is equal to zero (Figure 11.3)

Figure 11.3

2. The index of a closed curve which surrounds several singular points is equal to the sum of indices of these points.

3. The index of a closed trajectory (limit cycle or a trajectory near a center) is equal +1.

4. If at all points along the curve the phase vectors are directed inward (or outward), the Poincaré index of the curve is +1.

Now let us use this method in studying of local bifurcations on the phase plane. But first of all, what is a "bifurcation"?

The term bifurcation (from the latin word *bifurcus* meaning forked, split into two parts) means acquisition of a new quality by the motions of a dynamical system as a result of a small change of its parameters. In other words, bifurcation corresponds to a restructuring of the motion of a real system. Mathematically speaking, bifurcation is a change of topological structure of partitioning of the phase space of a dynamical systems into trajectories which is caused by small variation of system's parameters.

The last definition is based on the concept of the topological equivalence of dynamical systems. Two systems are topologically equivalent, i.e. have identical structures in phase space with respect to the partitioning into trajectories, if the motions of one of them can be transformed into motions of another via a continuous change of coordinates and time. Figure 11.4 shows a few examples of topologically identical phase planes.

A transition between different pictures on Figure 11.4 are not considered as bifurcations.

Now let us discuss briefly the most important bifurcations of equilibrium states on the phase plane:

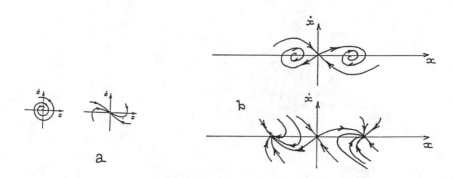

Figure 11.4

(1) The merging and subsequent disappearance of two equilibrium states.

The simplest example is the motion of a marble in a trough when a small bump on the bottom of the trough smooths out (Figure 11.5). In this case we speak about annihilation of a center and a saddle. You can easily check that the Poincaré index of curve N does not change and remains +1.

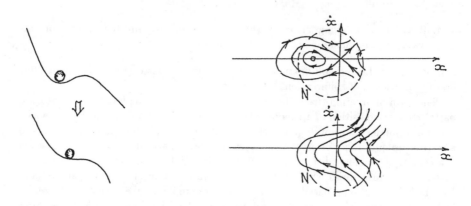

Figure 11.5

(2) We already studied the Andronov-Hopf bifurcation when a limit cycle appears from a focus (Figure 11.6). Figure 11.6b shows the moment of birth of the limit cycle which has zero amplitude.

(3) The birth of three equilibrium states from a single one: spontaneous symmetry breakdown. It is the opposite of the annihilation bifurcation, but there is a reason why it has its own name.

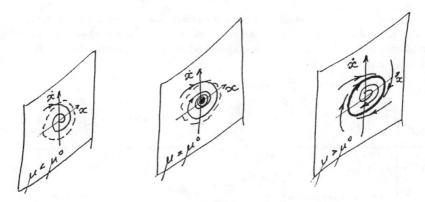

Figure 11.6

In each of cases 2 and 3 the Poincaré index of the curve does not change.

Look at Figure 11.7. A little bump appears on the bottom of the trough. On the phase plane one center changes into two centers and one saddle, thus providing conditions for existence of stable nonsymmetric motions in a completely symmetric system.

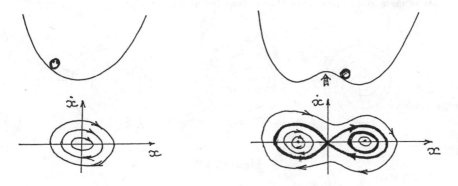

Figure 11.7

The local bifurcations of equilibrium states can be observed in the evolution of small perturbations in the system described by linearized equations. In the dynamical system $\frac{dx}{dt} = \mathbf{X}(x, \mu)$ (where $x(t)$ is the vector of physical variables, μ is a parameter). Small perturbations about the equilibrium state $x_0(\mu)$; $\mathbf{X}(x_0(\mu), \mu) = 0$ are described by $\xi(t) = x(t) - x_0(\mu)$, where

$$\frac{d\xi(t)}{dt} = \frac{\partial X(x_0(\mu), \mu)}{\partial x} \cdot \xi(t). \tag{48}$$

If $\xi(t) = \xi_0 e^{\lambda(\mu)t}$, then we have the usual characeristic equation for the roots $\lambda(\mu)$. If the roots do not lie on the imaginary axis of the complex plane (Figure 11.8), no bifurcation takes place in the neighborhood of the equilibrium state with a small parameter shift.

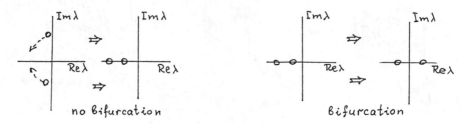

Figure 11.8

Only when one or several roots lie on the imaginary axis of the complex plane for μ equal to its critical value μ_c does the bifurcation occur. All the bifurcations of the disappearance or of the birth of equilibrium states correspond to the passage of one or several roots through the imaginary axis. An example is the birth of equilibrium state of node and saddle types that is depicted on Figure 11.9.

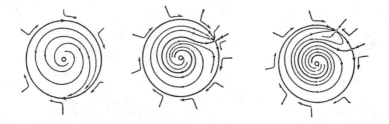

Figure 11.9

Such a bifurcation is encountered, for example, in the problem of mode competition (as in the previous lecture) or the competition of species feeding from one source (Figure 11.10).

The kinetic equations that describe the variation of the amplitudes of two competing modes or population of two competing species have the following form

$$\dot{X}_{1,2} = [1 - (X_{1,2} - \rho_{1,2} X_{2,1}) X_{1,2}]. \tag{49}$$

When both coupling coefficients $\rho_{1,2}$ are greater than 1, any species may win the struggle for existence, if the initial population of that species is large enough. As one of the parameters is decreased and becomes smaller than unity, only one of these species will survive under arbitrary initial conditions (Figure 11.10).

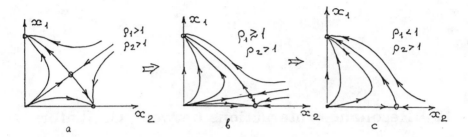

Figure 11.10

When two roots of a characteristic equation become purely imaginary, a limit cycle either appears from the equilibrium state or dies in it. This means that for all values of the parameter μ that are smaller (or larger) than the critical μ_c or rather close to it there exist a periodic solution that tends, when $\mu \to \mu_c$, to the time independent form $x_0(\mu)$. The stability of the limit cycle inherits the stability of the equilibrium state when $\mu = \mu_c$. This is already known to us as the Andronov-Hopf bifurcation.

12   Resonance Interactions between Oscillators

Let us start with a concrete example (Figure 12.1).

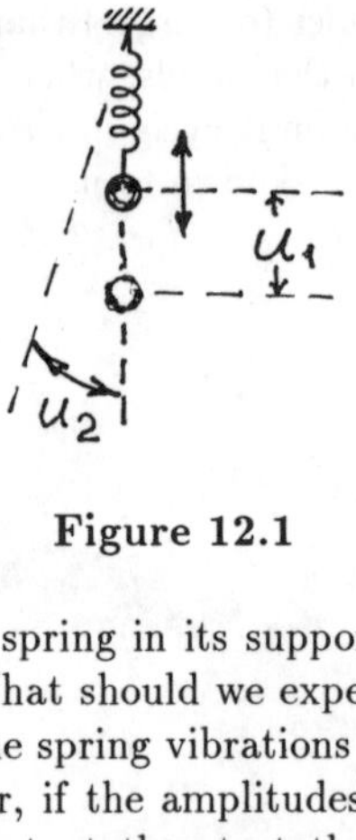

Figure 12.1

We have a pendulum with a spring in its support. This motion is confined to a plane in the gravitation field. What should we expect to see? If the amplitudes are very small, we may guess that the spring vibrations and the angular oscillations will be almost independent. However, if the amplitudes are large enough, the behavior is more interesting. Assume that at the start the spring is almost vertical with a very small angular velocity and is very stretched. We will see something like a parametric resonance. If the frequency of spring vibrations is twice that of the angular oscillations, the latter will absorb energy from the spring and its amplitude will increase. If we have the reverse initial conditions (the spring is at its equilibrium length and makes a large angle with the vertical line), it is clear that the amplitude of the spring vibration will also increase when the frequencies of each have the same ratio.

What are the most important features of this system? First of all there is no friction and therefore the energy is conserved and we may expect that when the amplitude of one oscillator is maximum the amplitude of the other is minimum. This means that if we start with a small amplitude of angular oscillations, they will grow at the expense of the energy stored in the spring vibrations. But this process must stop at least when the energy of these vibrations is almost completely drained. When this happens, the amplitude of angular oscillations is maximum and it will pump the energy back into the spring vibrations.

Therefore we can approximately plot the amplitudes as functions of time (Figure 12.2)

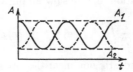

Figure 12.2

The second important fact is that the coupling between the spring and angular oscillations is nonlinear. One can easily obtain the equations of motion for this system:

$$\ddot{u}_1 + \frac{k}{m}u_1 = \ell(\dot{u}_2{}^2 - \frac{g}{2\ell}u_2^2)$$
$$\ddot{u}_2 + \frac{g}{\ell}u_2 = -\frac{1}{\ell}(\frac{g}{\ell}u_1 u_2 + 2\dot{u}_1 \dot{u}_2). \tag{50}$$

Notice that if the vertical frequency is twice that of the swing, there are terms in both right sides which are resonant. This is a very important and widely encounted phenomenon. The examples where this happens are coupled electrical circuits, coupled vibrations in large installations (buildings, engines etc.), coupled waves in plasma and on the surface of a fluid, thermal fluid convection and many other situations.

Now let us develop a mathematical method for the description of this system so that we could check our intuitive result shown in Figure 12.2. We will consider the case of a quadratic nonlinearity. Then all phenomena can be studied using the example of three oscillators with resonance coupling. In the absence of nonlinearity the system will oscillate in some normal modes $x_j; j = 1, 2, 3$ at some normal frequencies ω_j. In general we can write the following equations of motion for this system:

$$\ddot{x}_j + \omega_j^2 x_j = \mu f_j(x_1, x_2.x_3). \tag{51}$$

We are going to use the asymptotic method for the case $\mu \ll 1$, but we will look at a different interpretation of this method. The exact solution can always be written in the form

$$x_j(t) = (a_j(\mu t)e^{i\omega_j t} + cc) + \mu w_j(t). \tag{52}$$

The term in brackets looks familiar; it is the solution when $\mu = 0$. We can write the equation for the functions $w_j(t)$ using $\mu \ll 1$, so the equation is linear:

$$\ddot{w}_j + \omega_j^2 w_j = -2i\omega_j \dot{a}_j e^{i\omega_j t} + i2\omega_j \dot{a}_j^* e^{-i\omega_j t} + f_j(a_k e^{i\omega_k t} + cc) \, ; k = 1, 2, 3$$
$$= F(\mu t, t) \tag{53}$$

We know that if the function $F(\mu t, t)$ has resonance harmonics at frequencies $\approx \omega_j$ in its Fourier spectrum, the functions $w_j(t)$ will grow secularly as in a linear resonance. Therefore, the Fourier transform of $F(\mu t, t)$ these frequencies must be zero. This leads us to an equation for the slow amplitudes:

$$2i\omega_j \dot{a}_j = \frac{1}{T} \int_t^{t+T} f_j(a_k e^{i\omega_k t'} + cc) \, e^{-i\omega_j t'} \, dt'. \tag{54}$$

Since we consider only quadratic nonlinearities, we can write

$$f_j = \sum_{k,l} \alpha_{j,k,l} x_k x_l. \tag{55}$$

One can see that the integral is nonzero only if

$$\omega_1 + \omega_2 = \omega_3, \tag{56}$$

and this is what we call the `resonance condition`.

In the resonance case we can write the following equations for the complex slow amplitudes

$$\begin{aligned}
\dot{a}_1 &= -i\sigma_1 a_3 a_2^* \\
\dot{a}_2 &= -i\sigma_2 a_3 a_1^* \\
\dot{a}_3 &= -i\sigma_3 a_1 a_2.
\end{aligned} \tag{57}$$

Letting

$$a_j \to \frac{a_j}{\sigma_j} \, ; \; \sigma = \sqrt{\omega_1 \omega_2 \omega_3}. \tag{58}$$

we can reduce the equations of motion to

$$\begin{aligned}
\dot{a}_1 &= -i\sigma a_3 a_2^* \\
\dot{a}_2 &= -i\sigma a_3 a_1^* \\
\dot{a}_3 &= -i\sigma a_1 a_2.
\end{aligned} \tag{59}$$

Now we can say something about the behavior of this system. At first sight the system has 6 degrees of freedom: we have three equations for complex amplitudes. But luckily we can easily find a few integrals of motion. For doing this let us introduce the intensities of the modes $N_j(t) = |a_j|^2$.

From the equations for the N_j

$$\dot{N}_{1,2} = -i\sigma(a_3 a_2^* a_1^* - a_3^* a_2 a_1)$$
$$\dot{N}_3 = i\sigma(a_3 a_2^* a_1^* - a_3^* a_2 a_1) \tag{60}$$

it is easy to see that

$$\frac{d(N_1 - N_2)}{dt} = 0$$
$$\frac{d(N_3 + N_2)}{dt} = 0, \tag{61}$$

These are called the Manley-Rowe relations.

Since the modes are undamped, we may expect that our system has an energy integral. Since we know the energy of a simple oscillator is proportional to the intensity times its frequency. Let us multiply each of the equations for the N_j by ω_j and add them together. This leads to

$$\frac{d(\omega_1 N_1 + \omega_2 N_2 + \omega_3 N_3)}{dt} = i\sigma(a_3 a_2^* a_1^* - cc)(\omega_3 - \omega_1 - \omega_2), \tag{62}$$

which vanishes at resonance.

Our expression for the energy allows a very nice interpretation. As we remember, in quantum mechanics the energy of the j-th oscillator is given by $E = \hbar\omega_j N_j$, hence our definition of energy is the same as that for the quantum mechanical energy of an ensemble of oscillators. Because of this when people discuss the nonlinear interaction of oscillators (or waves), they sometimes speak of $N_{1,2,3}$ as of quanta of energy in each mode. In that sense the Manley-Rowe relations are just the rules of annihilation and creation of quanta.

One can prove that the Manley-Rowe relations and energy conservation provide three independent integrals of motion. Therefore, we can reduce the dimension of the phase space by three.

From energy conservation and the Manley-Rowe relations we can conclude something about how the system will behave starting from different initial conditions. First, let the intensity of the first mode (chosen to be the lowest frequency) be much larger than that of the others. Looking at the conservation laws we can say that the intensities can not change significantly (Figure 12.3).

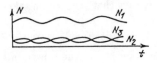

Figure 12.3

In the other extemem when the initial intensity of the highest frequency mode ω_3 is large nothing prohibits that energy from being alomost completely transferred into the other two modes and therefore we may expect to see something like Figure 12.4:

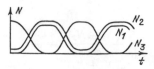

Figure 12.4

We can reduce the dimension of the phase space even further. To do so let as write the equations for the amplitudes and the phases $a_j = A_j e^{\Phi_j}$:

$$\dot{A}_1 = \sigma A_2 A_3 \sin \Phi$$
$$\dot{A}_2 = \sigma A_2 A_1 \sin \Phi$$
$$\dot{A}_3 = -\sigma A_2 A_1 \sin \Phi$$
$$\dot{\Phi} = -\sigma \left(\frac{A_1 A_2}{A_3} - \frac{A_1 A_3}{A_2} - \frac{A_2 A_3}{A_1} - 3 \right) \cos \Phi, \tag{63}$$

where $\Phi = \Phi_3 - \Phi_2 - \Phi_1$.

We see that only a certain combination of the phases enters the equations. Because of this we have a system of just four equations which, of course, has a four dimensional phase space. If we use the constraints due to the Manley-Rowe relations and energy conservation, we will end up with a one dimensional phase space. *This means that the system is integrable.* In this case we can find the solution in terms of elliptic functions, but this does not add much to the physical understanding of what is going on. So, instead of getting into this boring mathematics, let us consider a particular case when $\Phi = \frac{\pi}{2}$ and clearly does not depend on time. This statement is not as trivial as it seems because the equation for the phase has terms proportional to inverse of amplitudes. Since the amplitudes may get very close to zero, these terms can be very large and the derivative $\dot{\phi}$ will not be zero even if $\Phi \approx \frac{\pi}{2}$. If you study more carefully the behavior at the time when one of the amplitudes approaches zero, you will find that at this moment the phase suddenly changes from $+\frac{\pi}{2}$ to $-\frac{\pi}{2}$ or vice versa. To avoid complications associated with this fact we can simply allow the amplitudes to take negative values. This compensates the change of sign by $\sin \Phi$ in all equations.

The conservation of energy says that the motion takes place on the surface of the ellipsoid

$$\omega_1 A_1^2 + \omega_2 A_2^2 + \omega_3 A_3^2 = \text{constant.} \tag{64}$$

At the same time the motion is subject to the constraints of Manley-Rowe relations:

$$A_{1,2}^2 + A_3^2 = \text{constant} \; ; \; A_1^2 - A_2^2 = \text{constant.} \tag{65}$$

The intersections of all these surfaces are the phase trajectories we were looking for (Figure 12.5).

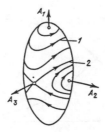

Figure 12.5

Now we can see that our intuition gave us the correct answers. Indeed, when the all the energy is initially in first or second modes (which both have low frequencies) there is not much energy exchange, and the oscillations of the amplitudes are very small (Figure 12.3). If the initial energy is mostly in the high-frequency mode, a great deal of it is transferred into the other modes, and the magnitude of amplitude oscillations is very large (Figure 12.4).

We can also see a rather unusual behavior. Assume that the initial amplitude of the third mode is zero and the amplitudes of two others are such that the phase point lies on the separatrix. Then, as we know, it will take a logaritmically infinite time for the system to come to the saddle point and it looks as if the energy is transferred irreversibly into the third mode (Figure 12.6).

Finally we note that the equations we have been studying are exactly the same as Euler's equations for a rigid body fixed at a point with differing moments of inertia I_j along its principal axes:

$$\dot{\Omega}_1 = \frac{I_2 - I_3}{I_1}\Omega_2\Omega_3$$

$$\dot{\Omega}_2 = \frac{I_1 - I_3}{I_2}\Omega_2\Omega_3$$

$$\dot{\Omega}_3 = \frac{I_1 - I_2}{I_3}\Omega_2\Omega_3, \tag{66}$$

where $I_3 \geq I_2 \geq I_1$, and the Ω_j are the angular velocities of the rigid body about the principal axes.

The phenomena discussed in this lecture are somewhat similar to the precession and nutation of a rigid body. You can think of other analogies between these two physical systems. For instance, the instability of the state with all energy in the high frequency mode corresponds to instability of rotation around the axis with the middle value of the moment of inertia and so on.

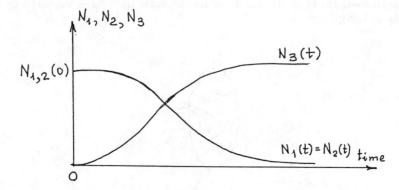

Figure 12.6

13　Solitons

We already know what will happen if we couple two or three oscillators to each other. Now we are about to find out how an ensemble of coupled oscillators behaves. This is a very difficult question because the behavior depends in general on very many factors: the kind of coupling, the kind of nonlinearity, geometry etc. As usual, we begin with the simplest model one can think of: a one dimensional chain of `linear` oscillators with nonlinear coupling between them (Figure 13.1).

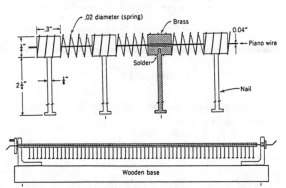

Figure 13.1

The chain is formed by identical pendula whose pivots are placed equidistant along a piano string. All pendula are connected to each other by springs which can twist, thus providing an additional torque.If we take $\theta_j(t)$ to be the angle the jth pendulum makes with the vertical, we can write the equation of motion for the system:

$$M\frac{d^2\theta_j(t)}{dt^2} = \kappa[\theta_{j+1} - 2\theta_j + \theta_{j-1}] - T\sin\theta_j. \tag{67}$$

This is a nonlinear differential difference equation. M is the moment of inertia of any pendulum, κ is the torque constant of the spring, and $T\sin\theta_j$ is the gravitational restoring torque.

Before studying the nonlinear dynamics of this system let us recall how it behaves at small θ_j. We can linearize the equations of motion to obtain

$$\ddot{\theta}_j + \omega_0^2\theta_j = \frac{\kappa}{M}(\theta_{j+1} - 2\theta_j + \theta_{j-1}), \tag{68}$$

where $\omega_0^2 = \frac{T}{M}$.

We can see that when the difference on the right side is zero at any j, the system oscillates at the frequency of an individual pendulum. It is clear why it happens: in this case all pendula oscillate in phase, the springs are at their equilibrium positions and do not provide any torque. What will happen if the phases are not the same? Suppose we pull one pendulum away from equilibrium when the others are hanging vertically. It is clear intuitively that the springs will force the neighbors away from the equilibrium positions. The neighbors in turn will drive their neighbors and so on.

This is shown in Figure 13.2:

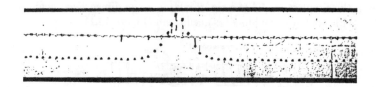

Figure 13.2

We will see waves propagating in the chain. Let us look for the solution in the form of single frequency oscillations:

$$\theta_j(t) = Ae^{i(\omega t - jka)} + \text{cc.} \tag{69}$$

The phase ωt is what we had before for a single oscillator except now ω is neither the eigenfrequency of the oscillator nor the frequency of a driving force. The quantity ka characterizes the phase shift between the jth pendulum and the $(j+1)^{th}$ pendulum. After we use this ansatz in the linear equations of motion, we arrive at the **dispersion relation** between the frequency ω and the wavenumber k:

$$\begin{aligned}
\omega^2 &= \omega_0^2 + \frac{2\kappa}{M}[1 - \cos ka] \\
&= \omega_0^2 + \frac{4\kappa}{M}\sin^2\left(\frac{ka}{2}\right).
\end{aligned} \tag{70}$$

The dispersion law for real ω and real k is shown in Figure 13.3.

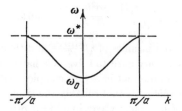

Figure 13.3

It is not hard to find a physical explanation for this dispersion relation. We already discussed the case when $k = 0$ and saw that the springs were at rest. If $k \neq 0$, there is an additional restoring torque due to the springs. The action of this torque is equivalent to an increase in the effective gravity for each pendulum, and therefore the frequency must also increase. Furthermore, higher values of frequency correspond to larger k because the larger k is, the stronger the force from the springs trying to move the pendula back toward the equilibrium position.

In the dispersion diagram we can see that in the frequency band $0 \leq \omega \leq \omega_0$, real k is not allowed. This means, for example, that if we start driving pendulum J at a frequency from this band the spatial dependence will be given by $e^{-a|k||j-J|}$. The excitation will not propagate away from the point where it first appears. The chain is not "transparent" to these frequencies. The frequency band where real k is permitted is called the transparency region. If we drive the chain at a frequency from this region, the solution will be a propagating wave. The excitation from one end of the chain can reach the other end without any loss of the energy. We can represent this solution as a sum (or integral) over sinusoidal propagating waves $\sin(\omega t \pm kaj)$ each of which travels left or right without changing form. These are the normal modes for the motion in our system. Any initial value or boundary value problem can be easily solved in term of these modes. This is all familiar from many other results in the study of linear systems.

Now let us see what the nonlinearity will do to our chain. Unfortunately, we are not used to dealing with difference differential equations. To simplify the problem let us consider the long wavelength limit, that is we assume that $\theta_j - \theta_{j+1} \ll \theta_j$. In other words, the variable θ does not change much in a distance a. More precisely we assume $ka \ll 1$.

In this case we can write

$$\theta_{j+1} - 2\theta_j + \theta_{j-1} \approx (\Delta x)^2 \frac{\partial^2 \theta(x,t)}{\partial x^2}. \tag{71}$$

In this approximation the original equation is equivalent to the partial differential equation[10]

$$\frac{\partial^2 \theta(x,t)}{\partial x^2} - \frac{\partial^2 \theta(x,t)}{\partial t^2} = \sin \theta, \tag{72}$$

where distances are measured in units of $\sqrt{\frac{\kappa (\Delta x)^2}{T}}$, and time in units of $\sqrt{\frac{M}{T}}$.

This last equation is still very hard to solve. We have a system with a continuous spectrum of modes; they can interact in a very complex way, and it is a real problem to describe the motion in this system. Probably the situation would simplify a lot if we could find the analog of sinusoidal waves in a linear problem. Maybe the nonlinear system also has some normal modes which maintain their form propagating in space. Of course, these modes will no longer be sinusoidal waves. What we definitely know about them is that since they preserve their form, the function $\theta(x,t)$ can depend only on the combination $\xi = t \pm \frac{x}{v}$, where v is some velocity.

Using this reduced dependence of $\theta(x,t) = \theta(\xi)$ on its independent variables, we arrive at a wave equation we can solve:

$$\frac{d^2 \theta(\xi)}{d\xi^2} = \frac{v^2 \sin \theta}{v^2 - 1}. \tag{73}$$

This is the familiar nonlinear pendulum equation. The only difference is that now we have a parameter v which is the velocity of the steady wave solution. The velocity may be greater or smaller than 1. What difference does it make? Figure 13.4 shows the phase space for each cases. They are the same up to a shift of π along the axis. From the physical point of view these two cases correspond to different initial states of the chain. If $v < 1$ all pendula are initially in their stable position. If $v < 1$ the pendula begin in the unstable position.

We can integrate our Sine-Gordon equation once to obtain

$$\frac{d\theta(\xi)}{d\xi} = \pm \sqrt{\frac{2v^2(E - \cos \theta)}{v^2 - 1}}, \tag{74}$$

where E is a constant of integration.

When we studied nonlinear oscillators, we realized that it was hard to find a general solution for such equations of motion. We definitely can find the solutions of linearized equations, if we consider small deviations from the stable equilibrium states. This means that the sinusoidal waves we discussed just a moment ago correspond to motions near the collection of centers which are the equilibrium state in the phase space of the Sine-Gordon equation. Luckily, sometimes we can also find the solution

---

[10] In this form of a partial differential equation it is known as the Sine-Gordon equation which is a (bad) pun on the linearized version known in relativistic quantum theory as the Klein-Gordon equation. The Sine-Gordon equation also appears in the discussion of Josephson junctions in low temperature physics.

for the motion along the separatrix. In the case at hand the solutions for such separatrix motions, occurring at $E = \pm 1$, are

$$\theta(x - vt) = 4 \arctan[\exp\{\pm(\frac{x - vt}{\sqrt{1 - v^2}})\}], \tag{75}$$

for $E = 1, v < 1$, and

$$\theta(x - vt) = 4 \arctan[\exp\{\pm(\frac{x - vt}{\sqrt{1 - v^2}})\}] + \pi \tag{76}$$

for $E = -1, v > 1$.

These two functions are shown in Figure 13.4.

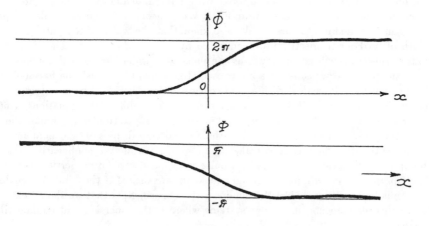

Figure 13.4

These functions do not look like solitary pulses. But we should not forget that the states shifted by $\pi, 2\pi, 4\pi$, etc are the same because θ is a phase. Therefore, if we look at a snapshot of the wave solutions we will see something like Figure 13.5:

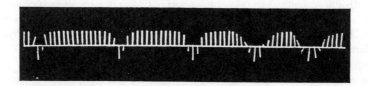

Figure 13.5

In the first part of Figure 13.4, the soliton (or solitary wave or solitary pulse) is a localized region of a fixed shape where the pendula rotate by the angle 2π over the whole length of the system. For this solution $\theta(x,t) = 0$ at $x = \pm\infty$. This is called a "simple soliton". The second part of this Figure shows a different behavior. As $x \to \pm\infty$ the pendula go to their unstable positions. This soliton is called a "dark soliton". It is clearly unstable; perturbations will send it into the first case. See Figure 13.5.

This is a very nice looking phenomenon. But is it worth studying? Perhaps we can observe solitons only in some very special and unrealistic situations. Let us discuss the mechanisms which lead to appearance of solitons.

Can we observe the propagation of pulses with the fixed form in linear systems? Yes, we can, but only in a dispersionless medium. If the dispersion law is not linear, different harmonics (sinusoidal waves with different ω and k) will propagate with different velocities. As we know from linear wave theory, this causes a pulse propagating in this medium to spread. We also know that the width of the pulse and the width of its Fourier spectrum are connected by $(\Delta x)^2(\Delta k)^2 \geq 2\pi$. Therefore, the Fourier spectrum will get narrower as the pulse gets wider. What will nonlinearity do to the pulse? If we consider a linear oscillator, we will have just one harmonic at the frequency ω. If we deal with a nonlinear oscillator, there will be infinitely many harmonics. The same will happen in the case of partial differential equations. Nonlinearity generates higher harmonics, that is widens the spectrum and, therefore, may make the pulse more narrow. Thus, in the general case dispersion and nonlinearity compete: the dispersion tries to make the pulse wider while the nonlinearity acts in the opposite direction. As a result, a pulse with quite specific properties survives. These properties (width, height, speed and so on) are such that the action of the dispersion is balanced by that of the nonlinearity. This suggests that we should expect to see this phenomenon in various systems where both dispersion and nonlinearity coexist.

Historically, the first equation where solutions in form of solitary waves were found was the Korteweg-de-Vries (KdV) equation:

$$\frac{\partial u(x,t)}{\partial t} + u(x,t)\frac{\partial u(x,t)}{\partial x} + \beta\frac{\partial^3 u(x,t)}{\partial x^3} = 0. \tag{77}$$

It was introduced in 1895 to describe long gravity waves on a surface of water, and since then it has been used as a classical example of nonlinear wave equations. This equation is very simple, and at the same time it is very general. Of course, we could derive it from the fluid equations of motion. But we would like to show that in a medium with small dispersion and a quadratic nonlinearity, the KdV equation provides a quite general description regardless of the specific physical system.

If we have a linear wave propagating in one direction (from left to right, for example) in a dispersionless medium, it will satisfy

$$\frac{\partial u(x,t)}{\partial t} + c\frac{\partial u(x,t)}{\partial x} = 0. \tag{78}$$

What will change if we have small dispersion? As we know a dispersion law for a homogeneous and isotropic medium always can be written in the form $\omega^2 = f(k^2)$, because there should not be any difference between waves propagating in different directions. For a **dispersionless medium** we can write this law in the form $\omega^2 = c^2 k^2$. If we assume that the dispersion is very small, we can expand $f(k^2)$ in Taylor series $\omega^2 = c^2 k^2 + \Gamma k^4$. If now we solve for $\omega(k)$ using $\Gamma k^2 \ll c^2$ [this what we mean by `small` dispersion], we will find that $\omega(k) = \pm ck \pm (-\beta)k^3$. For the case of normal dispersion $\Gamma < 0$, and $\beta > 0$. Then for waves travelling from left to right we can write $\omega(k) = ck - \beta k^3$. The corresponding partial differential equation will be

$$\frac{\partial u(x,t)}{\partial t} + c\frac{\partial u(x,t)}{\partial x} + \beta\frac{\partial^3 u(x,t)}{\partial x^3} = 0. \tag{79}$$

If we introduce a quadratic nonlinearity in the most general case, we will have the following equation:

$$\frac{\partial u}{\partial t} + c\frac{\partial u}{\partial x} + \beta\frac{\partial^3 u}{\partial x^3} + \alpha u^2 + \delta u\frac{\partial u}{\partial t} + \gamma u\frac{\partial u}{\partial x} = 0. \tag{80}$$

We can write the solution as the sum $U(x,t) + v(t)$ where $v(t)$ is a solution of

$$\dot{v} + \alpha v^2 + \delta v\dot{v} = 0. \tag{81}$$

We are interested in the solution which has $u \to 0$ as $x \to \pm\infty$, and, therefore, $v(t) \equiv 0$. Now we have the equation

$$\frac{\partial u}{\partial t} + (c + \delta u)\frac{\partial u}{\partial x} + \beta\frac{\partial^3 u}{\partial x^3} = 0. \tag{82}$$

From this equation we can see the physical origin of the nonlinearity. The nonlinear term reflects the fact that the speed of wave propagation in a nonlinear medium depends on the amplitude of the wave. Finally, we can go into the system moving with the velocity c and send $u \to \frac{u}{\delta}$. This brings us to the KdV equation.

Now let us look for the solution to the KdV equation in the form of a steady wave: $u(x,t) = u(\xi = x - Vt)$. Recalling our treatment of the Sine-Gordon equation we obtain

$$(V - u)\frac{du}{d\xi} + \beta\frac{d^3 u(\xi)}{d\xi^3} = 0, \tag{83}$$

or after integration $\beta\frac{d^2 u}{d\xi^2} - (Vu - \frac{u^2}{2}) = $ constant. We can eliminate the constant on the right hand side by the transformation $u \to u +$ another constant. The soliton solution is represented again by the separatrix solution in the phase plane (Figure 13.6). Using the result from an earlier lecture we can display the solitary pulse:

$$u(x,t) = u_{\max} \cosh^{-2}[\frac{x - Vt}{\Delta}], \tag{84}$$

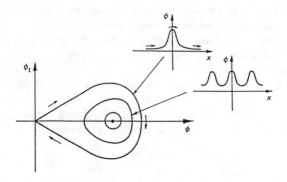

Figure 13.6

where $4\beta = V\Delta^2$ and $12\beta = \Delta^2 u_{max}$.

u_{max} is the height of the pulse, V is its velocity, and Δ, the characteristic width (Figure 13.7).

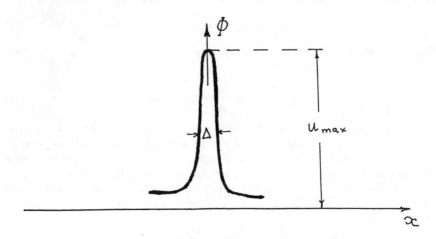

Figure 13.7

Thus, we have found:

1. $12\beta = \Delta^2 u_{max}$. This means that the higher the soliton, the narrower it is.

2. $\Delta^2 V = 4\beta$; $U_{max} = 3V$. That is, the wider (and shorter) the soliton, the slower it will travel.

It is very important that the speed, the height, and the width of the KdV soliton are uniquely related. This means that we have one parameter family of solutions. Figure 13.8 shows the dependence of the phase portrait on the parameter V.

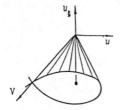

Figure 13.8

There is still one unanswered question. What if this solution is unstable? Then we would never observe solitons in real life, because any infinitesimal disturbance would destroy them. It turns out that solitons are often stable with respect to not only perturbations of the profile of the wave, but also with respect to small changes in the original equation. We can even introduce a small amount of dissipation, and the soliton will stay alive for a very long time.

Solitons remain stable and regain their form even when they interact with each other. Let us see what happens when a fast soliton passes a slow one (Figure 13.9).

When the pulses overlap significantly they move at approximately the same speed. During this time the larger soliton becomes smaller and the smaller one grows. Correspondingly, the fast soliton slows down and the slow one speeds up. The collision of solitons looks approximately like that of elastic balls! Isn't this amazing! As the fast soliton runs away from the slow one the picture looks exactly like before except that the solitons changed their relative positions. What is most important is that the form remained unchanged.

Now, let us look more attentively at Figure 13.9c. This pulse is not a one-soliton solution. As time goes on it breaks into two. This gives us a very good hint about that we will see if we consider the initial value problem. Assume at the initial time we have a pulse of arbitrary form travelling from left to right. Obviously, the pulse will change form.

It turns out that it breaks into a chain of solitons (Figure 13.10).

To summarize these facts, solitons maintain their form when propagating and colliding, and an arbitrary pulse breaks into a sequence of solitons with different parameters. All this means that solitons in a nonlinear medium play the same role as sinusoidal waves play in a linear one. They can be considered as the "eigenwaves" for a nonlinear problem.

This result may seem unbelievable. The nonlinear system has infinite number of degrees of freedom and at the same time exhibits such a simple behavior. Why do

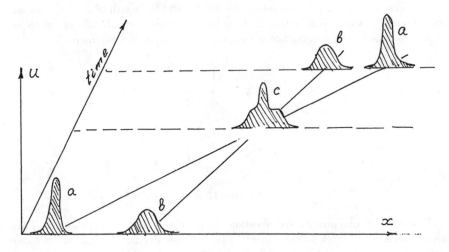

Figure 13.9

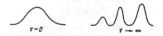

Figure 13.10

not we see anything like chaos? We know that any system will exhibit no complex (chaotic) behavior if we can find as many constants of motion as degrees of freedom. In our present case we have an infinite number of degrees of freedom and ... an infinite number of constants of motion!

To illustrate this let us introduce a function w (Miura et. al., 1968) by

$$u = w + \epsilon \frac{\partial w}{\partial x} + \epsilon^2 w^2, \tag{85}$$

where ϵ is an arbitrary constant. It is very convenient to write the KdV equation in the form

$$\frac{\partial u}{\partial t} + 6u \frac{\partial u}{\partial x} + \frac{\partial^3 u}{\partial x^3} = 0, \tag{86}$$

which we can always do by an appropriate scaling.

Using the definition of w, we arrive at

$$\frac{\partial w}{\partial t} + \frac{\partial}{\partial x}[\frac{\partial^2 w}{\partial x^2} - 3w^2 - 2\epsilon w^3] = 0,$$

$$\frac{\partial w}{\partial t} + \frac{\partial F_w}{\partial x} = 0. \tag{87}$$

for all ϵ. This is in the form of a conservation law with F_w the flux of w. But ϵ is an arbitrary constant, and therefore, we have an infinite number of conservation laws for functions of the original variable u. To see these we solve for w(u) as a power series in ϵ:

$$w = w_0 + \epsilon w_1 + \epsilon^2 w_2 + \cdots \tag{88}$$

We find $w_0 = u$, $w_1 = -\frac{\partial u}{\partial x}$, $w_2 = \frac{\partial^2 u}{\partial x^2} - u^2$, and so on. The conservation equation can now be integrated over x from $-\infty$ to $+\infty$, and assuming the solution goes to zero at the end points, we arrive at

$$\frac{dI_n}{dt} = 0$$

$$I_n = \int_{-\infty}^{+\infty} w_n(u(x,t))\, dx. \tag{89}$$

Since $n = 0, 1, 2, \ldots$, we indeed have an infinite number of conservation laws. Furthermore, it is easy to see that all conserved quantities (except the first two) have terms with powers of u larger than unity, and, therefore, the constants of motion should be independent. A detailed argument shows this to be true.

14   Steady Propagation of Shock Waves

Some of you probably noticed that in the previous lecture we considered a rather idealized situation. We neglected dissipation, but that is present in any physical system. Does this mean that everything we said is useless? By no means! We pointed out that solitons are often stable even with respect to small perturbations in the equations we use to describe them. Therefore, if at some time we have a soliton going in some direction, then when a small amount of dissipation is introduced, we will see the same "soliton" but with slowly decreasing amplitude and speed. However, we still have no clear idea about what will happen if, in addition to the dissipation, some energy is pumped into the system, perhaps from the outside world through the boundaries of the system. Of course, there is such a variety of ways to do this that we could spend hours discussing them. Instead let us consider the simplest and, probably, the most important situation when the solution to the driven, dissipative problem is zero at $+\infty$ (or at $-\infty$) and has some finite value at $-\infty$ (or at $+\infty$). We will again try to find a solution in the form $u(x,t) = u(x - vt)$. If we succeed, we will find a description of 'shock' waves in dissipative media.

Now we put the words 'shock' in quotes since usually engineers or hydrodynamicists talk about shock waves in media without dissipation and focuses on sharp changes in pressure coming from a supersonic jet or an explosion. We use the term, perhaps slightly loosely, to mean a front propagating with uniform speed where some interesting physical quantity changes rapidly. This includes fluid dynamical systems as well electromagnetic or other waves.

In other words we will be interested in the situation shown in Figure 14.1:

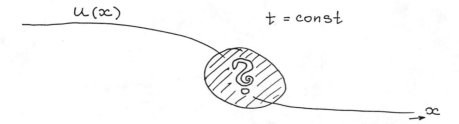

Figure 14.1

What happens in the transition region? Let us find out. But first of all we need an equation to work with. The KdV equation as shown in the previous lecture is the general equation for a system with small quadratic nonlinearity and small dispersion. Therefore, it is sensible to take this equation and add a term responsible for dissipation. This leads to what we call the KdV-Burgers equation:

$$\frac{\partial u(x,t)}{\partial t} + u\frac{\partial u(x,t)}{\partial x} + \beta\frac{\partial^3 u(x,t)}{\partial x^3} = \nu\frac{\partial^2 u(x,t)}{\partial x^2}. \tag{90}$$

The simplest way to see that the term $\nu\frac{\partial^2 u(x,t)}{\partial x^2}$ is indeed responsible for dissipation is to consider the linearized equation and find the dispersion relation. We then see that when $\nu > 0$, the imaginary part of the frequency corresponding to a real wave vector will be $i\nu k^2$, and this leads to damping since $u(x,t) \sim e^{i\omega t} = e^{-\nu k^2 t + i\mathrm{Re}\omega t}$.

We are looking for a solution in the form of a steady propagating wave that is $u(x,t) = u(\xi = x - Vt)$. This leads to the ordinary differential equation

$$-V\frac{du(\xi)}{d\xi} + u\frac{du(\xi)}{d\xi} + \beta\frac{d^3u(\xi)}{d\xi^3} = \nu\frac{d^2u(\xi)}{d\xi^2}, \tag{91}$$

and after a first integration we arrive at

$$\beta\frac{d^2u}{d\xi^2} - \nu\frac{du}{d\xi} + \left(\frac{u^2}{2} - Vu\right) = \text{constant.} \tag{92}$$

As we saw in the previous lecture, we can set the constant to zero.

Now this is the equation of motion for a particle with mass β in the potential $W(u) = -V\frac{u^2}{2} + \frac{u^3}{6}$ with friction constant ν.

This is shown in Figure 14.2.

Solutions of the original equation are represented by the usual trajectories in phase space. We are looking for quite a specific solution for which the function $u(\xi)$ at $\xi \to \pm\infty$ approaches a constant value. If this is to be so, the trajectory corresponding

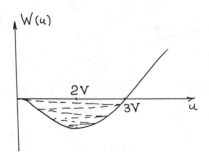

Figure 14.2

to this solution must go from one (unstable) equilibrium state to another one. In the phase plane of our system we have exactly two equilibrium states: at $u = 0$ and at $u = 2V$. The one at zero is obviously a saddle (the top of a hill). The second one may be either an unstable focus or node depending on how strong the dissipation is. As $\xi \to +\infty$ (that is, $t \to -\infty$ at a fixed location or $x \to +\infty$ at a fixed time) the trajectory must approach zero along the stable separatrix. At $\xi \to -\infty$ ($t \to +\infty$ or $x \to -\infty$), the system must be at the focus. Thus the boundary condition lets us pick the unique trajectory. Note that in this case (unlike the initial value problem considered in the previous lecture), we do not have any freedom in the choice of the constant V. This is because we are given that $u(x \to -\infty) = 2V$.

We can consider two extreme cases corresponding first to small and then to large dissipation. If the dissipation is very small and the dispersion is very large, the second equilibrium state will be a focus and the trajectory leaving this state will make very many oscillations before it reaches the saddle.

See Figure 14.3:

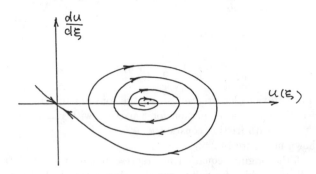

Figure 14.3

The profile of this wave is shown in Figure 14.4:

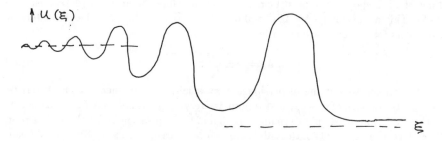

Figure 14.4

As in the case of solitons in the KdV equation, the phase portrait of the system depends on a single parameter V which is defined by the boundary condition. We can show this dependence in three dimensional space (Figure 14.5).

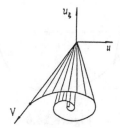

Figure 14.5

Although the leading edge of the wave in Figure 14.4 looks like a chain of solitons with different parameters, these "solitons" propagate at the same speed in the dissipative medium. Of course, they spend some energy fighting with the friction force but they keep getting energy from the infinitely long tail of the wave. This is an interesting phenomenon but can be hardly called a shock wave. It is pretty obvious what is responsible for the absence of a steep step; namely, the dispersion which is always trying to make any sharp function wider. Therefore, if we want to see a shock wave we better consider a medium with a very small dispersion. Let us see how small it should be.

We are interested in the time it takes the system to pass from one equilibrium state to another. To get an idea about how this time depends on the parameters in our equation let us consider the characteristic equations near the equilibrium states:

$$\beta\lambda^2 - \nu\lambda + (u_{\text{equilib}} - V) = 0. \tag{93}$$

First of all, a shock wave must be a step, and therefore the unstable state must be a node. This means that $\nu^2 - 4\beta V > 0$ or $\nu > 2\sqrt{\beta V}$. In this case the characteristic exponents for the node are given by

$$2\beta\lambda_\pm = \nu \pm \sqrt{\nu^2 - 4\beta V}. \tag{94}$$

We can see that if the dispersion is very small, $\beta \to 0$, then the characteristic exponents are both of order $\frac{\nu}{\beta}$, that is both are very large. When looking for exponents of the saddle, we need only to change the sign in the expression under the square root. Therefore, the exponents of the saddle are also very large, when $\nu \ll 4\beta V$. This means that the transition will be very fast. The phase diagram will look like Figure 14.6.

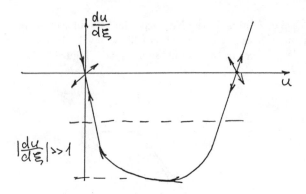

Figure 14.6

The corresponding shape of the shock wave is shown here

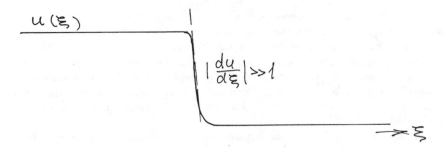

Figure 14.7

Thus, if we have a propagating transition from one value of a physical variable to another depending on the parameters we can see different structures of the transition. If the dispersion is very large compared to other parameters, we will see a bunch of solitons (Figure 14.4). If the dispersion is negligible, then the variable will experience a sharp jump. In intermediate cases we will see the structures shown in Figure 14.8.

Propagation for intermediate dispersion:

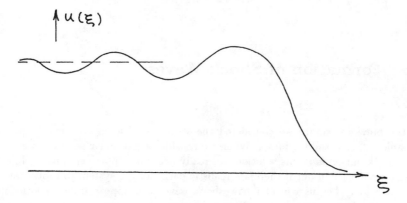

Figure 14.8

15 Formation of Shock Waves

In the previous lecture we considered the structure of a region where a physical variable suffers a sudden jump. When we consider a particular solution, we always should ask ourselves if this solution can really occur in a physical system. In other words, we must find out if there are realistic paths for the system to evolve towards that solution. Let us consider how shock waves can appear from smooth initial conditions. As an example we will consider gravitational waves on the surface of a body of water (Figure 15.1). We will assume that the wavelength is much larger than the depth ($\lambda >> h_0$); that is, the water is shallow. We consider only two spatial dimensions.

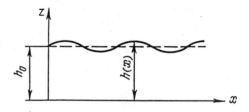

Figure 15.1

For long waves we can assume that the velocity is directed along the x-axis and does not depend on the z-coordinate (except in a narrow layer near the bottom). If we neglect dissipation, we can write the Euler equation for the x-component of the velocity using Newton's law:

$$\rho\left(\frac{\partial v(x,t)}{\partial t} + v\frac{\partial v(x,t)}{\partial x}\right) = -\frac{\partial p(x,t)}{\partial x}, \tag{95}$$

with the pressure $p(x,t) = p_0 + \rho g(h(x,t) - h_0)$. p_0 and h_0 are the undisturbed pressure and height. We also consider the density to be constant. The conservation of mass in this case takes the form

$$\frac{\partial h(x,t)}{\partial t} + \frac{\partial(v(x,t)h(x,t))}{\partial x} = 0. \tag{96}$$

This pair of equations is known as the shallow water equations, and they appear widely in the study of thin layers of fluid as in the oceans and atmospheres of the earth.

If we linearize this system near the equilibrium values v_0 and h_0, we find the dispersion relation $\omega = k(v_0 \pm \sqrt{gh})$, so the system has no dispersion. This suggests that the connection between h and v must be algebraic: $h = h(v)$. Assuming this relation our equations become

$$\frac{\partial v(x,t)}{\partial t} + v\frac{\partial v(x,t)}{\partial x}) = -g(\frac{dh}{dv})\frac{\partial v(x,t)}{\partial x}$$
$$\frac{dh}{dv}[(\frac{\partial v(x,t)}{\partial t} + v\frac{\partial v(x,t)}{\partial x}] = -h\frac{\partial v}{\partial x}. \tag{97}$$

Since we started with equations for v and h but linked them by our assumption that h is a function of v only, we expect to find a consistency condition for this ansatz. It is

$$(\frac{dh(v)}{dv})^2 = \frac{h(v)}{g}, \tag{98}$$

and implies that

$$\sqrt{h(v)} = \frac{v}{2\sqrt{g}} + \text{constant}. \tag{99}$$

The equation for v(x,t) is then

$$\frac{\partial v}{\partial t} + (v \pm \sqrt{gh(v)})\frac{\partial v}{\partial x} = 0. \tag{100}$$

Generalizing from this we may write the equation for waves propagating in one direction as

$$\frac{\partial v}{\partial t} + V(v)\frac{\partial v}{\partial x} = 0. \tag{101}$$

One can check that the solution of this equation can be written in the form $v(x,t) = v(x - V(v)t)$. This looks like a steady propagating wave solution, but it is not. The problem is that the solution is not written in explicit form. V is expressed as a function of v which is a function of time and space. At this stage we can not find v(x,t) explicitly. Yet the implicit form of the solution lets us make substantial progress.

If we consider the function $\Psi(v)$ which is the inverse of $v(x - V(v)t)$, that is, $\Psi(v) = x - V(v)t$. Take the derivative with respect to x of this expression, we find

$$1 - t\left(\frac{dV(v)}{dv}\right)\left(\frac{\partial v}{\partial x}\right) = \left(\frac{d\Psi(v)}{dv}\right)\left(\frac{\partial v}{\partial x}\right). \tag{102}$$

Time enters this equation as a parameter, and, therefore, if we know $\frac{d\Psi(v)}{dv}$, we have a ordinary differential equation for $v(x,t)$. If we are given the function $v(x,0)$, then we can conclude that

$$\frac{d\Psi(v)}{dv} = \left(\frac{\partial v}{\partial x}\right)_{t=0}^{-1}. \tag{103}$$

Using this one can solve the equation for $\frac{\partial v}{\partial x}$ for a general initial condition. Let us consider, instead, the evolution of $\frac{\partial v}{\partial x}$ in the case of surface waves on shallow water. In general

$$\left(\frac{\partial v(x,t)}{\partial x}\right)^{-1} = \left(\frac{\partial v(x,0)}{\partial x}\right)^{-1} + t\frac{dV(v)}{dv}, \tag{104}$$

and $V(v) = v + \sqrt{gh(v)}$. One easily finds that $\frac{dV}{dv} = 3/2$ for the shallow water equations.

Let us describe qualitatively how the wave shown in Figure 15.2a will evolve. From $\frac{dV}{dv} = 3/2$ one can see that the part when $\frac{\partial v}{\partial x} > 0$, becomes smoother, that is, the derivative with respect to the spatial coordinate decreases. But when $\frac{\partial v}{\partial x} < 0$, the derivative increases with the absolute value remaining negative (Figure 15.2b). When $t = t^* = \frac{2}{3}|\left(\frac{\partial v(x,0)}{\partial x}\right)^{-1}|$, the derivative becomes infinite (Figure 15.2c), and after that the wave breaks (Figure 15.2d).

This is known as the gradient catastrophe. This is what happens when a wave hits a shallow region near the shore.

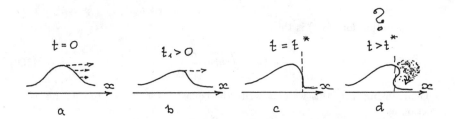

Figure 15.2

After the gradient catastrophe occurs a shock wave may form, and then dissipation smooths the function, and the solution remains singlevalued. Thus, a shock wave can appear in an initial value problem. In some systems, however, multivalued solutions can evolve beyond the gradient catastrophe point. This happens with shallow water waves and waves in modulated particle beams.

16 Solitons. Shock Waves. Wave Interaction. The Spectral Approach

We already used the spectral approach to understand the actions of nonlinearity and dispersion, but only a qualitative analysis was presented. Can we use the spectral method for a quantitative description of nonlinear continuous systems? Yes, we can. But the nonlinear spectral method is very different from what we do in linear systems. Specifically, in a nonlinear system different spectral components do not, as a rule, evolve independently. Harmonics with different wave numbers and different frequencies interact with each other. Therefore, we would expect that in this case the equations of evolution for spectral amplitudes will in general be differential or even integral equations and not algebraic as in linear systems. It is a very hard problem to find these equations, but as usual one can often do this in the case of a *weak nonlinearity*. In this case we can again use the averaging method with a slight change.

We can write the approximate solution $u(\mathbf{x}, t)$ to whatever equations we are considering in the form

$$u(\mathbf{x}, t) = \sum_{j=1}^{M} a_j(\mu t, \mu \mathbf{x}) e^{i(\mathbf{k}_j \mathbf{x} - \omega_j t)}, \tag{105}$$

where μ is a small number, and the frequency and the wave vector are connected by the dispersion relation $\omega_j(\mathbf{k}_j)$. We use this form of $u(\mathbf{x}, t)$ in the equations of motion, and we then average over time and space to find equations for the complex amplitudes a_j. This problem is similar to that of the interaction of oscillators with different frequencies. Again one can show that if we have a quadratic nonlinearity and strong dispersion, the elementary processes are two or three wave interactions. We know that the interaction will be significant only if the resonance conditions are satisfied. In our case these conditions must be written both for the frequencies and the wave vectors:

$$\omega(\mathbf{k}_1) + \omega(\mathbf{k}_2) \quad = \quad \omega(\mathbf{k}_3)$$
$$\text{and}$$
$$\mathbf{k}_1 + \mathbf{k}_2 \quad = \quad \mathbf{k}_3. \tag{106}$$

Figure 16.1 presents an example of a resonance diagram for the case of one-dimensional waves.

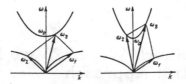

Figure 16.1

In practice these resonance conditions need not be satisfied exactly. The point is that since the amplitudes change slowly in time and space, the spectrum of each interacting wave is not really a delta function. Rather it has a narrow peak with its center at that frequency and that wave vector. Therefore, the resonance conditions become less restrictive:

$$\omega(\mathbf{k}_1) + \omega(\mathbf{k}_2) \quad = \quad \omega(\mathbf{k}_3) + \Delta\omega$$
$$\text{and}$$
$$\mathbf{k}_1 + \mathbf{k}_2 \quad = \quad \mathbf{k}_3 + \Delta\mathbf{k}. \tag{107}$$

Now, in this lecture we will only consider the case of a homogeneous one dimensional medium with amplitude

$$u(\mathbf{x}, t) = \sum_j a_j(\mu t) e^{i(\mathbf{k}_j \cdot \mathbf{x} - \omega_j t)} + \text{c.c.}, \tag{108}$$

and, therefore, $\Delta\mathbf{k} = 0$.

Let us see what this means for the case of the KdV-Burgers equation.

$$\frac{\partial u}{\partial t} + u\frac{\partial u}{\partial x} + \beta\frac{\partial^3 u}{\partial x^3} = \nu\frac{\partial^2}{\partial x^2}. \tag{109}$$

First, assume that both dispersion and dissipation are absent and the initial condition is a sinusoidal wave with wave number k and frequency ω_0. Since there is no dispersion, the resonance conditions will be satisfied exactly for any triplet $n\omega_0, m\omega_0, (n + m)\omega_0$ where n and m are integers.

This we show in Figure 16.2:

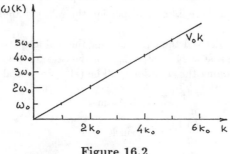

Figure 16.2

This means that the energy will be transferred from the initial periodic excitation to harmonics at $n\omega_0$. If ν and β were zero, this process would last forever, and as time goes to infinity the energy in each harmonic would be close to zero because the total energy is finite and would be distributed among an infinite number of harmonics. Of course, this will never occur because at time t^*, the gradient catastrophe takes place and after that either the spectral method does not work, because the field becomes multi-valued, or we need to take into account dispersion $\beta \neq 0$ or dissipation $\nu \neq 0$ or both.

If ν is non-zero but very small, it is easy to see that the harmonics with higher wave numbers are damped more strongly. Therefore the spectrum after some time will almost stabilize. "Almost" because the energy must decrease and so must the intensities of the harmonics. This "almost" time independent spectrum corresponds to a slowly decaying series of shocks where the nonlinearity transfers some energy into higher harmonics, but this energy is immediately dissipated.

The situation looks more unusual if the damping is zero but the dispersion is not. In this case as the dispersion line bends more and more, the resonance condition is satisfied less and less precisely (Figure 16.3):

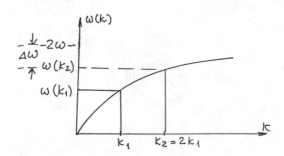

Figure 16.3

Therefore the energy transfer stops at the frequency $\omega(\mathbf{k}_3)$ for which $\omega(\mathbf{k}_1) + \omega(\mathbf{k}_2) - \omega(\mathbf{k}_3) > \Delta\omega$. To understand why this happens let us consider the case when $\mathbf{k}_1 = \mathbf{k}_2 = \frac{\mathbf{k}_3}{2}$. Since there is some dispersion the relation among the frequencies is $\omega_1 = \omega_2; \omega_3 = 2\omega_1 + \Delta\omega$.

We have already derived the equations for the real amplitudes of the two waves and for the phase shift $\Phi = 2\Phi_1 - \Phi_2 - \Delta\omega t$ for the case $\Delta\omega \neq 0$. You can easily check that for general detuning the equations will be (after some scaling transformations)

$$\frac{dA_1}{dt} = -A_1 A_2 \sin\Phi$$
$$\frac{dA_2}{dt} = A_1^2 \sin\Phi$$
$$\frac{d\Phi}{dt} = -\left(2A_2 - \frac{A_1^2}{A_2}\right)\cos\Phi - \Delta\omega \tag{110}$$

There is an energy integral $A_0^2 = A_1^2 + A_2^2$. Let us consider the behavior of the system on the phase plane.

First introduce new variables $X = A_2 \sin\Phi$ and $Y = A_2 \cos\Phi$. Then our system can be rewritten in the form

$$\dot{X} = A_0^2 - (X^2 + Y^2) - 2Y^2 - \Delta\omega Y$$
$$\dot{Y} = 2XY + \Delta\omega X. \tag{111}$$

The phase plane of this system for different values of detuning is shown in Figure 16.4:

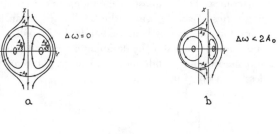

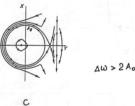

Figure 16.4

If the detuning is small enough ($\Delta\omega < 2A_0$), we can see that there is strong energy exchange between the modes. If we start with almost all energy in the first mode, our initial condition for A_2 is close to zero. As you can see in Figure 16.4(a,b) the system leaves the state $(A_1, A_2) = (A_0, 0)$ and goes to the state $(0, A_0)$. Since the final state is a saddle point the energy transfer will be irreversible (Figure 16.5(a)). If the initial state is not exactly on the x-axis, we will see oscillations (Figure 16.5(b)). If the detuning is large ($\Delta\omega > 2A_0$), the initial state of interest is in the neighborhood of the center (Figure 16.4(c)). Therefore, we will see oscillations of very small amplitude (Figure 16.5(c)). In other words, there will be no any significant energy exchange. In this frequency region the transformation of the energy into higher harmonics stops.

Now we understand the role of dispersion in a dissipationless medium $\nu = 0$. When we have only two modes, we can see either strong energy exchange between the modes, for small $\Delta\omega$, or fast phase oscillations with the amplitudes being constant, for large $\Delta\omega$.

In the general case, when there are many interacting modes, we expect to see three different kinds of behavior depending on the parameters in the system and the initial conditions:

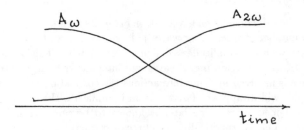

Figure 16.5 a

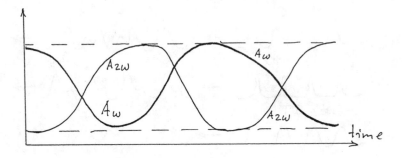

Figure 16.5 b

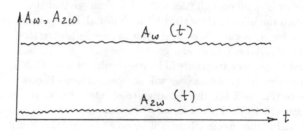

Figure 16.5 c

1. When the dispersion is small enough, namely the dispersion line is almost linear, we expect to see the formation of stationary waves such as solitons or solitons chains (cnoidal waves);

2. If the initial conditions are not so special we will see a periodic energy exchange not only between two harmonics, but among many of them and the initial harmonic waves (see Figure 16.6).

3. Finally, when the dispersion is very strong (the dispersion line is very nonlinear) and the number of interacting waves is large, we expect to see very fast phase variations with all amplitudes changing very slowly. This is because in the case of strong dispersion many waves will interact with large detunings. Since each wave is involved in many interactions the correlation among the phases in each triplet will disappear. This problem is similar to the many body problem in classical mechanics. The later is known to be nonintegrable, that is the bodies may behave chaotically. In our case we have a wave system which is not integrable. This is the problem of weak wave turbulence which will be considered in the next lecture.

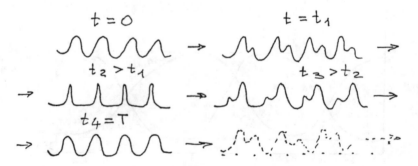

Figure 16.6

17 Weak Turbulence.
Random Phase Approximation

So far we observed only regular motions in systems with large numbers of degrees of freedom. This regularity is closely related to the fact that in those systems all modes are synchronized. In general this is not true, and in some cases the energy is transferred among the modes in a very twisted and unpredictable way. The field fluctuates randomly, and we observe what we call turbulence. How can we possibly describe this phenomenon?

Let us first understand what causes the complexity. In all previous lectures we considered interaction between two or among three waves as though they were the only waves in the universe. More precisely, we assumed that each wave is involved in interaction only within one triplet. In general, of course, this is not true. There may be very many triplets involving the same mode. As we mentioned before, exact synchronization is very hard to arrange. Synchronization with small detuning is more likely to occur, but it will involve very few modes. Therefore, if a mode is involved in very many interactions this implies that the detuning is very large. In terms of the two wave interaction considered in the previous lecture $\Delta\omega \gg 2A_0$. This means that the equation for the phase will have a very large coefficient on the right side: $\frac{d\Phi}{dt} \approx \Delta\omega$. Thus the rate of the phase variations is much larger than that of amplitudes. Furthermore, the interaction with so many waves may make the phase fluctuate randomly. These two facts give us hope that we may average everything over the phases and find the (slow) equations for the amplitudes or for the numbers of excitations in different modes.

Let us apply this outlook to the three-wave interaction. The basic equations are

$$
\begin{aligned}
\dot{a}_1 &= \sigma a_3 a_2^* \\
\dot{a}_2 &= \sigma a_3 a_1^* \\
\dot{a}_3 &= -\sigma a_1 a_2
\end{aligned}
\tag{112}
$$

95

In the spirit of what we have said we are looking for a solution in the form

$$a_j(t, \mu t) = N_j^{\frac{1}{2}} e^{i\Phi_j(t)} + \mu a_j'(t, \mu t), \tag{113}$$

where μ is a small parameter. If we insert these expressions into the equations for the intensities $|a_j(t)|^2$:

$$
\begin{aligned}
\dot{N}_1 &= \sigma a_1^* a_2^* a_3 + c.c. \\
\dot{N}_2 &= \sigma a_1^* a_2^* a_3 + c.c. \\
\dot{N}_3 &= \sigma a_1 a_2 a_3^* + c.c.,
\end{aligned}
\tag{114}
$$

we will have for the terms proportional to μ

$$
\begin{aligned}
\mu a_{1,2}' &= \sigma (N_{2,1} N_3)^{\frac{1}{2}} \int_{t_0}^t e^{i(\Phi_3 - \Phi_{2,1})} dt' \\
\mu a_3' &= \sigma (N_1 N_2)^{\frac{1}{2}} \int_{t_0}^t e^{i(\Phi_1 - \Phi_2)} dt'.
\end{aligned}
\tag{115}
$$

The terms on the right hand sides are indeed very small. If the phases were absolutely uncorrelated the integrals would be zero. If we assume that the phases are random but have finite correlation time Figure 17.1, the integrals will be small if this time is small.

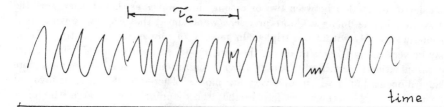

Figure 17.1

Using the last equation we can find the equations for $N_j(t)$. For instance for N_3 we will have

$$
\begin{aligned}
\dot{N}_3 = \sigma^2 \Big[& N_1 N_2 \Big\langle e^{i(\Phi_1 + \Phi_2)} \int_{t_0}^t e^{-i(\Phi_1 + \Phi_2)} dt' \Big\rangle \\
& - N_3 N_1 \Big\langle e^{i(\Phi_3 - \Phi_1)} \int_{t_0}^t e^{-i(\Phi_3 - \Phi_1)} dt' \Big\rangle - N_2 N_3 \Big\langle e^{i(\Phi_3 - \Phi_2)} \int_{t_0}^t e^{-i(\Phi_3 - \Phi_2)} dt' \Big\rangle \Big]
\end{aligned}
\tag{116}
$$

The notation $< \bullet >$ corresponds to an average over the phases. After the averaging is performed the expressions $< \bullet >$ become constant, and, therefore, we arrive at the following equations.

$$
\begin{aligned}
\dot{N}_3 &= \xi(N_1 N_2 - N_2 N_3 - N_1 N_3) \\
\dot{N}_{1,2} &= -\xi(N_1 N_2 - N_2 N_3 - N_1 N_2).
\end{aligned} \tag{117}
$$

For good reason these equations are called the kinetic equations for waves. They look exactly like the equations one can derive for the concentrations of components of a chemical reaction away from chemical equilibrium.

We can see that the intensities in our equation still satisfy the Manley-Rowe relations: $N_1 - N_2 = \text{constant}, N_3 + N_{1,2} = \text{constant} = K_{1,2}$. Using these relations, we can express $N_{1,2}$ in terms of N_3. Then for N_3 we will have the following equation:

$$
\dot{N}_3 = \xi \left[K_1 K_2 - 2(K_1 + K_2)N_3 + 3N_3^2 \right]. \tag{118}
$$

This is a gradient equation, and it can be written in the form

$$
\dot{N}_3 = -\frac{\partial U}{\partial N_3} \tag{119}
$$

with

$$
U(N_3) = -K_1 K_2 N_3 + (K_1 + K_2)N_3^2 - N_3^3. \tag{120}
$$

Therefore the phase space of this system is a line Figure 17.2. The equilibrium states can be easily found:

$$
N_3^0 = \frac{K_1 + K_2}{3} \pm \sqrt{\frac{(K_1 + K_2)^2}{9} - \frac{K_1 K_2}{9}}. \tag{121}
$$

As we know, the equations for the intensities always evolves towards a stable equilibrium state and after that stays in that state Figures 17.3. Using energy conservation and the Manley-Rowe relations, one finds that the stable equilibrium states for all three modes can be written in the form $N_{1,2,3} = \frac{\text{constant}}{\omega_{1,2,3}}$. This is the law of equipartition from thermodynamics which says that the energy at equilibrium is equally distributed among the different degrees of freedom. In this case we can think of the constant as the temperature of the system. The important point is that the energy exchange is irreversible in time. In this system the irreversibility appears not because of dissipation but because of the assumption that the phases are random.

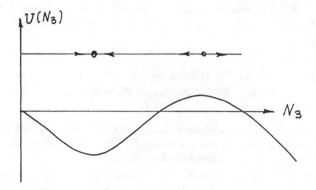

Figure 17.2

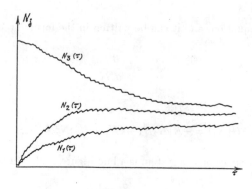

Figure 17.3

Now we must make the final step and generalize the intensity equations for the case when each wave is involved in interactions with many triplets:

$$\dot{N}_{\mathbf{k}} = \int \left[|\sigma_{\mathbf{k},\mathbf{k}',\mathbf{k}''}|^2 (N_{\mathbf{k}'}N_{\mathbf{k}''} - N_{\mathbf{k}}N_{\mathbf{k}'} - N_{\mathbf{k}}N_{\mathbf{k}''}) \delta(\mathbf{k} - \mathbf{k}' - \mathbf{k}'') \right] d\mathbf{k}' d\mathbf{k}'' = I_c\{N_{\mathbf{k}}\}. \tag{122}$$

There is again a solution corresponding to thermodynamic equilibrium: $N_{\mathbf{k}} = \frac{T}{\omega_{\mathbf{k}}}$
This is in Figure 17.4.

This equation was derived for a medium without dissipation, and, therefore, should be corrected. If we include the terms responsible for dissipation all motions will decay to zero as time goes to infinity unless there is some energy supply. Therefore, a more general form of this equation is

$$\dot{N}_{\mathbf{k}} = I_c\{N_{\mathbf{k}}\} + D(\mathbf{k}, N_{\mathbf{k}}) - \Gamma(\mathbf{k}, N_{\mathbf{k}}),\tag{123}$$

where I_c is the term describing wave interactions, Γ is the dissipation function and D is the term describing the distribution of the energy supply over the spectrum.

This is a very important result. This equation describes weak wave turbulence in a variety of systems: waves in plasmas, gravitational surface waves on a fluid, nonlinear optics and others, where we can assume that the phases change randomly.

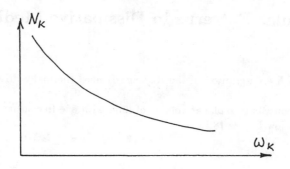

Figure 17.4

18   Regular Patterns in Dissipative Media

Let us start with a description of the Rayleigh-Bénard thermally driven convection experiment.

The setup consists of a planar volume of fluid with one free surface and uniform heating from below Figure 18.1.

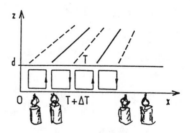

Figure 18.1

When the temperature difference is less than some critical value, one observes heat transfer from bottom to top by heat conduction which is governed by thermal diffusion. There is no macroscopic motion of the fluid, and the energy is carried from the bottom to the top by individual molecules. However, when the temperature gradient reaches some critical value, the fluid starts moving. The system can transfer the heat more efficiently by this than continuing to conduct the heat by molecular motions. The fluid forms rolls as seen in Figure 18.1. If we consider a horizontally infinite fluid layer, the size of the rolls will be determined by the local parameters of the problem. In a large box the orientation of the rolls depends on the initial conditions.

How can we explain this phenomenon? We can consider small deviations from the conducting solution in terms of sinusoidal waves whose amplitudes are proportional to $e^{\Gamma t}$. This is suggested by the fact that the coefficients in the differential equations of motion are constant in time. Since these deviations are very small, we can write linear equations describing their evolution. We will find that for each mode there is a critical temperature difference when Γ becomes positive. It is convenient to express this in terms of the dimensionless Rayleigh number $Ra \propto \Delta T$ to characterize the temperature difference. One can compute the critical Rayleigh number for waves with different wave vectors.

This function is shown in Figure 18.2.

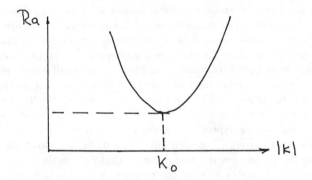

Figure 18.2

As we can see, when the temperature difference reaches its critical value, two counterpropagating waves become unstable. The nonlinear terms stabilize their amplitudes at some final value, and they form a standing wave which corresponds to convective rolls.

In the general case we can write the expression for the temperature or fluid velocity in the form

$$\phi(\mathbf{r}, t) = \sum_{j=1}^{N} a_j(t) e^{i(\mathbf{k}_j \mathbf{r} - \omega_j t)} + c.c. \tag{124}$$

Here N is the number of excited modes.

If we want to understand why and how different patterns appear, we need equations describing the evolution of different modes. But it is very hard to derive these equation from hydrodynamics. Besides we know that a similar experiment can be performed with many other systems, and, therefore, there must be a way to derive the equations using very general arguments. First of all, we know that in the linear approximation the amplitudes a_j must decay or grow exponentially. Therefore, we know the first term in the equation of motion:

$$\frac{da_j}{dt} = \Gamma_j a_j + \dots \tag{125}$$

Secondly, we know that this growth must saturate at some value due to the presence of nonlinear terms. Therefore, we can write the second term:

$$\frac{da_j}{dt} = \Gamma_j a_j - \alpha_j a_j |a_j|^2 + \dots \tag{126}$$

If we take into account possible nonresonant interactions among the waves, we will arrive at the following equation

$$\frac{da_j}{dt} = \Gamma_j a_j - \alpha_j a_j \left(|a_j|^2 + \sum_{l \neq j} \rho_{jl} |a_l|^2 \right) \tag{127}$$

When $\rho_{jl} = \rho_{lj}$, this is a gradient system; therefore, it has only fixed points in its phase space. The system will evolve to one of them and will stay there.

The coupling parameters ρ_{jl} depend on the overlap integral of modes in phase space: the more the overlap, the stronger the interaction. If we consider two waves with orthogonal wave vectors the interaction between them will be very weak. Because of this we can observe square lattices in some experiments: parametrically excited capillary waves, for example. How will the modes with significant overlap interact with each other? The answer is not trivial. The modes draw energy from the same source and, therefore, we expect to see mode competition. But in this case there may be a few modes competing, and the phase space will not look that simple. Depending on the parameters the equilibrium state to which the system will relax to may correspond to a two- or four- or six- or eight- or even a many-mode state. The number of modes should be even because if the system allows a right propagating wave it also allows a left propagating wave. These join to form a standing wave. In the case of two-mode state, we have rolls with an arbitrary orientation. Should the pattern be always so nice and ordered? No, there are examples when it is not! What do you think you will see if the pattern is formed by eight waves with the wave vectors shown in Figure 18.3a? The solution for this problem is very unexpected Figure 18.3b. These waves form a quasi-crystal. This structure is ordered and irregular at the same time. It is ordered because it is formed by a finite number of modes. But if you compute the spatial correlation function it will decay along any direction, and will be close to unity only for large scales See Figure 18.3c. Fascinating, isn't it?

Now let us come back to the differential equation for $a_j(t)$. We did not include the terms describing resonant wave interaction. For simplicity, we would like to consider only three-wave interactions. We know that this kind of interaction is possible only if we have a quadratic nonlinearity. Physically this corresponds, for example, to a temperature dependent viscosity coefficient in a fluid convection. If we introduce a quadratic nonlinearity in our equation we will have

$$\frac{da_j}{dt} = \Gamma_j a_j - \alpha_j a_j \left(|a_j|^2 + \sum_{l \neq j} \rho_{jl} |a_l|^2 \right) - \sum \beta_{lq} a_q^* a_l^*. \tag{128}$$

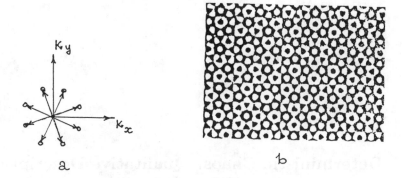

Figure 18.3

The second sum is taken over all harmonics which satisfy the resonance condition: $k_j + k_q + k_l = 0$. Let us consider only three such modes. If the system is close to its threshold, all three wave vectors must be of the same length. Therefore we can describe the resonance condition by the diagram in Figure 18.4a.

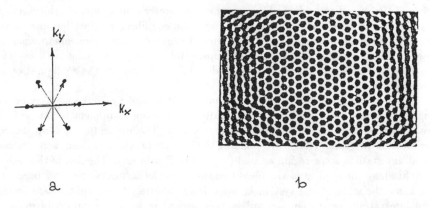

Figure 18.4

If, at the initial time, we have waves 1 and 2, they will pump wave 3 to larger amplitude. This means that if the coupling coefficient is large enough, all three standing waves will grow simultaneously. This leads to the formation of a honeycomb structure Figure 18.4b. In this case our system has the only stable fixed points corresponding to the coexistence of the three modes with equal amplitudes. The critical value of the coupling coefficient can be derived from our equations, if there are only three interacting modes.

19 Deterministic Chaos. Qualitative Description

Before discussing the origin of chaos in dynamical systems, let us list a few relevant facts about the behavior of phase trajectories in one and two dimensional phase space. In one dimensional phase space we have only trivial equilibrium states, that is stable and unstable fixed points. All motions are very simple. The system evolves towards one of the stable points, and having reached this point it stays there. As we go from one to two dimensional phase space we encounter a new type of trajectory: a limit cycle. The behavior of the system near the equilibrium points becomes more complicated and diverse. We know how to use the characteristic exponents to separate the equilibrium states into different groups. But there is one point we have not discussed yet. We did not study how predictability depends on the configuration of the phase plane.

What do we mean when we speak about good or bad predictions. Assume we make a measurement at some moment of time. Since our equipment is not perfect, we know the state of the system with finite precision. This means that if, for instance, we measure the parameters corresponding to the state (x, y), the system actually can be at any state in some region around (x, y) as in Figure 19.1. The size of the region is defined by the quality of the measurement. Now let ask ourselves how precisely we know the state of the system at some later time relying on that measurement and integrating the system forward in time according to the equations of motion. Since we have some spread in the initial conditions for this integration, we will have some spread in the states reached at that later time. We say that we make a good prediction, if the spread in the states we reach at that later time is less than or of the same order as the spread we begin with. Otherwise we say that the prediction is bad.

Let us illustrate this with a few examples. First consider a stable focus. If we follow the evolution of a "drop" of initial conditions, we will see that the size of the drop will decrease and as time goes to infinity the drop will draw itself into a point at the origin Figure 19.2a. This means that we have good prediction for the motions

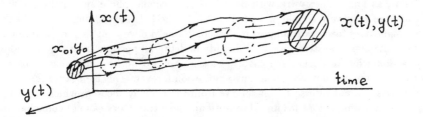

Figure 19.1

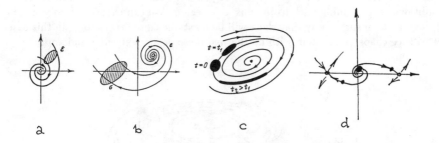

Figure 19.2

in this system. In the case of an unstable focus, if we consider the same drop, we will find that it expands as time goes on. We can not make an accurate prediction. From the practical point of view we have a very strange situation: we know the equations of motion exactly, we know the initial state very accurately, but if we wait long enough we cannot say precisely how the system evolves Figure 19.2b.

You may say: "Wait a moment! The situation is not that dramatic. Why do we consider such an unphysical example? If we have only an unstable focus on the phase plane the trajectories and all physical variables go to infinity and this is impossible in any real system." This is indeed correct. In practice we always have some region of the phase plane where all trajectories end up. This means that in two dimensional phase space we always have some, perhaps somewhat limited, predictability. For example, if the unstable focus is inside of a limit cycle, we can predict the amplitude and the frequency of the final motion with very high precision and only the phase of these oscillations is hard to predict Figure 19.2c. If we consider an unstable focus and a pair of stable points in the attracting region of the phase plane, we can always predict the long time behavior of the system with probability not less than $\frac{1}{2}$ Figure 19.2d.

Thus, as long as we work with only two dimensional phase space, the possible motions are relatively simple and predictable. Nonetheless, studying motion in the phase plane does teach us an important lesson. Namely, if we want to find the origin of chaos in dynamical systems, we should study in detail all possible **unstable** motions. The reason why we do not see chaotic motions in two dimensional phase space is that the system can not stay in an unstable region forever. This is because the phase trajectories can not intersect [this follows from the Cauchy theorem which says that the solution is unique for each initial condition], and a trajectory can not come back to the vicinity of an unstable equilibrium point after it leaves that point. If we consider, however, three dimensional phase space, such behavior becomes possible! Look at Figure 19.3. A trajectory leaves the unstable focus and at some time leaves the plane of the focus and jumps back to the region close to the origin. A drop of initial conditions keeps coming back to the unstable region. We can choose the parameters of the system in such a way that there will be no stable closed trajectory. In this case we will see chaos in a system with no thermal noise or any external random force.

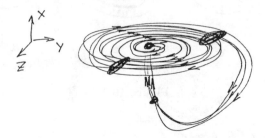

Figure 19.3

Let us study what happens to a drop of initial conditions Figure 19.4. First of all as the system evolves in the unstable region, the drop is getting stretched. If there is no dissipation, the volume of the drop must be constant according to Liouville's theorem, if there is dissipation the volume of the drop must decrease in time. Therefore the drop can be stretched in one direction but then definitely must be compressed in some other direction. Then, since we want all motion to remain bounded, we must bend the drop a little bit so that its size does not significantly exceed the original size in any direction. After that we repeat this operation all over again. This transformation is similar to the so-called baker transformation. A baker takes a piece of dough, rolls it out and puts some spice in just one place. Then she folds the dough and rolls it out again. The dough is stretched when the baker rolls it, this is instability. The folding corresponds to mixing of trajectories. After many foldings and rollings, the spice is homogeneously distributed throughout the dough. This is what will happen with our drop of initial conditions. As time goes to infinity the drop spreads over a finite region of phase space and gets mixed. This means that if we make a measurement at some time with a very good but finite accuracy, there will be a time after which we will not know how the system is going to evolve. At some moment we will have no accurate idea about where the system is in the phase space.

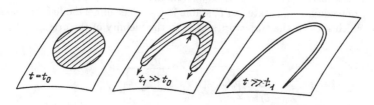

Figure 19.4

You may wonder how one could construct an attracting region, if all motions inside the region are unstable. This is an excellent and puzzling question. It turns out that this is possible. The trick is that a trajectory can be unstable in one direction and stable in the another as in a saddle point. For instance, if we have two unstable limit cycles, we can, in principle,arrange the phase space in such a way that the trajectory which just left one limit cycle is immediately attracted by the other instead of going to infinity. After it does a few oscillations about this limit cycle it leaves it only to be caught by the first.

See Figure 19.5.

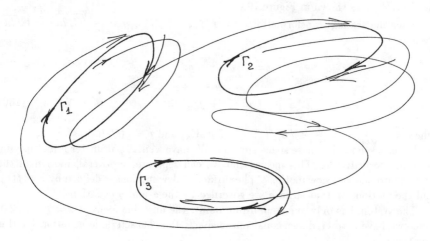

Figure 19.5

Now let us consider a real physical system which exhibits chaotic behavior. In the electrical circuit shown in Figure 19.6 we have a negative resistance element. This can be accomplished by using an vacuum tube or a transistor. We also have a tunnel diode as a nonlinear element. C_1 is the spurious capacity of the tunnel diode. $C_1 \ll C$. The negative resistor provides the instability of trajectories. This is very easy to see because, if we remove the tunnel diode, the phase space of the system will have only one equilibrium state: an unstable focus. The tunnel diode is the nonlinear element which returns the trajectories back to the vicinity of the origin.

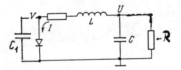

Figure 19.6

One can derive the following equation of motion for the circuit:

$$L\frac{dI}{dt} = RI + (U - V)$$

$$C\frac{dU}{dt} = -I$$

$$C_1\frac{dV}{dt} = I - I_{td}(V) \tag{129}$$

Here $I_{td}(V)$ is as shown in Figure 19.6a.

After the scaling transformations $x = I/I_m, z = V/V_m, y = \frac{U}{I_m}\sqrt{C/L}, \tau = t\sqrt{LC}$ we have

$$\dot{x} = 2hx + y - gz$$

$$\dot{y} = -x$$

$$\mu\dot{z} = x - f(z) \tag{130}$$

where $h = \frac{1}{\sqrt{LC}}, g = \frac{V_m}{I_m}\sqrt{C/L}, f(z) = \frac{1}{I_m}I_{td}(zV_m)$ and $\mu = g(C_1/C)$.

Let us see how the phase space appears. We have a small parameter μ multiplying the time derivative $\dot{z}$. This means that we can separate the overall motion in the system into fast and slow motions. The surface of slow motions is defined by $x = f(z)$, and the motions on that surface are governed by the x and y equations.

The system has only one equilibrium point: the unstable focus at $x = y = z = 0$. Since we have a three dimensional phase space, the characteristic equation for this point will be

$$\lambda^3 + p\lambda^2 + q\lambda + r = 0 \tag{131}$$

If we choose the parameters of our system properly, $\text{Re}\lambda_{1,2} > 0$ and $\text{Re}\lambda_3 < 0$. This equilibrium state is called a saddle focus Figure 19.7.

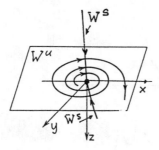

Figure 19.7

Using our knowledge of the surface of slow motions and the kind of equilibrium state, we can easily construct the phase space of this system Figure 19.8.

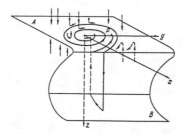

Figure 19.8

Let us speculate what the time series looks like. If we start from some point close to the origin, first we will see linearly growing oscillations (region A, Figures 19.8 and 19.9. Then the system reaches the edge of the surface of slow motions and jumps from it. This jump happens very rapidly, and after that the system goes slowly through region B. As the system reaches the second edge of the slow motion surface, the fast motion flow brings it back to the region close to the origin. Then everything happens again. Under some circumstances the system starts each time from a different point in the neighborhood of $(0,0,0)$, and roughly speaking the number of oscillations it will make before falling of the edge will be different each time.

The resulting time series shown in Figure 19.9 has all the properties of a random process: the autocorrelation function decays rapidly, the signal has a wide band

Fourier power spectrum, etc. All this despite that fact it has been generated by a dynamical system whose evolution is governed by deterministic differential equations with no random external force.

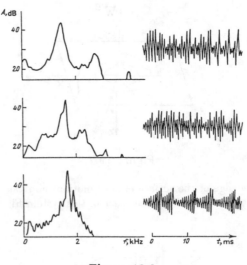

Figure 19.9

20 Description of a Circuit with Chaos. Chaos in Maps

As we saw in the previous lecture, a dynamical system may exhibit chaotic behavior. For a physicist the problem is to describe this behavior somehow and to find how its characteristics depend on the parameters of the system. In general this is a very hard problem and well beyond our present understanding of the dynamcis of nonlinear systems. However, in the system which we considered in the previous lecture we had a small parameter multiplying the time derivative, and we were able to separate the motion into fast and slow parts. After that the system looked simpler. Now we will try to simplify it even further by constructing a Poincaré section.

Let us consider the map formed by the intersections of a phase trajectory with the secant plane ($x = 0, y > 0$) shown in Figure 20.1, and let us find the map induced by the flow in going from the Poincaré section to itself: $y(j + 1) = F(y(j))$.

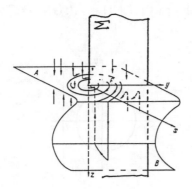

Figure 20.1

One can assume that the diode characteristic $f(z)$ looks like Figure 20.2; that is, it has a linear part which we call A where $f(z) = \frac{z}{\alpha}$ for $z < \alpha$, and a nonlinear section we call B:

This is illustrated here

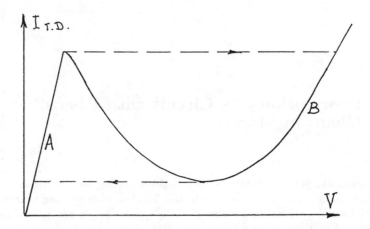

Figure 20.2

The function $F(y(j))$ must consist of two parts: the first one corresponds to an unwinding of the trajectory while the voltage across the diode is in part A of the IV characteristic, and the second one corresponds to the motion in part B, which returns the trajectory back in the neighborhood of the origin. For the first part we obviously have

$$\ddot{x} - 2\nu\dot{x} + x = 0; \quad \dot{y} = x, \tag{132}$$

where $\nu = h - \frac{\alpha g}{2}$. For this part $y(j+1) = F_1(y(j)) = e^{2\pi\nu}y(j) \equiv k y_j$, as shown in Figure 20.3.

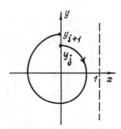

Figure 20.3

When $y(j) > y_{\text{edge}}$, the pair of vertical fast motions and the slow motion in part B of the IV characteristic take the trajectory back to almost where it started. Therefore, this part of $F(y(j))$ must look like Figure 20.4.

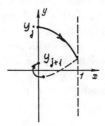

Figure 20.4

When we put these two pieces together, we will arrive at the map describing all motions in the system Figure 20.5.

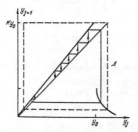

Figure 20.5

It is easy to make sure that all trajectories enter the region inside the boundary of the attractor. This is ensured by the shape of $F_2(y(j))$. At the same time if k is large enough all motions inside this region will be unstable: $|\frac{dy(j+1)}{dy(j)}| > 0$. You can play with this map, and you will convince yourself that there are no closed trajectories which are stable and that motions on this attractor generate a 'random' looking sequence of numbers.

Note that the map preserved all basic properties of the continuous system which are important for existence of chaos. If we consider an interval of initial states, we will find that as the discrete time goes on it is stretched and bent, stretched and bent ... If time is large enough, the states coming from the original interval will be scattered in a larger interval, which is called the size of the attractor, according to some distribution function.

To illustrate how this happens let us consider a very simple map: $x(k) = [2x(k-1)]$. The brackets correspond to taking the fractional part of the expression enclosed. The diagram for this map is shown in Figure 20.6a. If you follow some trajectory, you will probably find the behavior very complicated. At the same time one can show that there are infinitely many periodic trajectories in this system. To illustrate this we consider the maps over two and three steps as shown in Figures 20.6b and c. We see that as we increase the number of iterations, the number of fixed points also increases and therefore there are closed trajectories with periods 1, 2, 3, and so on. But it is also easy to see that all these periodic motions are unstable. As time goes by, the system must evolve towards a stable trajectory. But since all limit cycles and fixed points are unstable the only possibility left is that this trajectory will correspond to chaotic motion which never repeats itself.

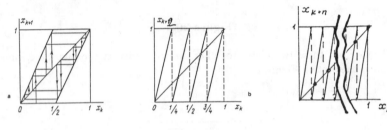

Figure 20.6

It is very easy to understand why this happens using binary numbers. If the initial state happens to be a rational number less than one but greater than zero it can be written as $0.(*)(*)(*)(*)(*)(*)(*)....$ where * is some combination of 1's and 0's. For instance, $0.101110111011101110111011....$ is a rational number equal to 11/15. Multiplication by two and subtraction of the integer part corresponds to a shift of the decimal point to the right and placing zero to the left of the new position of the decimal point. So, if we iterate the number we used as an example we will be getting:

```
0.011101110111011101110
0.111011101110111011101
0.110111011101110111011
0.101110111011101110111
0.011101110111011101110
```

Clearly we have a periodic motion with period 4. If we choose an irrational number as the initial condition, we will have at each step new irrational number because no group of 1's and 0's appears in the initial number periodically. Since the measure of irrational numbers is larger than that of rational numbers, if we draw an initial state we will pick, with probability unity, an irrational number. Therefore in practice we will see only nonperiodic motions.

Now, what can we say about the physical parameters of the system when it wanders in a chaotic region? We can not predict the exact state of the system. The only description we can develop is a statistical one. The statistical ensemble in this case is the ensemble of initial conditions. One can write the equations of evolution for the initial state probability density. In our case this equation is very simple:

$$\rho_{j+1}(F(x)) = \sum_{l=1,2} \rho_j(x) \frac{1}{|\frac{dF(x)}{dx}|_l} \tag{133}$$

where the summation is taken over both branches of $F(x)$.

If you follow the evolution of some initial distribution of initial states, you will easily convince yourself that at each step this distribution smooths more and more. As time goes to infinity the distribution approaches $\rho_j(x) = \frac{1}{2}$. This happens with the distribution of initial states in any chaotic system. In some sense the distribution function behaves in the same way as the dynamical variable in a gradient system. As time approaches infinity the distribution function approaches some limit. This limit is called the invariant measure and can be found from the condition $\rho_j(x) = \rho_{j+1}(x) = P(x)$. In the case at hand, we can prove that the invariant measure is $\frac{1}{2}$.

$$P(F(x)) = \sum_{l=1,2} P(x) \frac{1}{|\frac{dF}{dx}|_l} = \frac{1}{2} \sum_{l=1,2} P(x). \tag{134}$$

We see that $P(x) = $ constant satisfies this equation. The normalization condition requires $P(x) = \frac{1}{2}$. Using this probability density we can find any moment of $x(j)$. For instance, $< x > = \frac{1}{2}$, and this is very reasonable.

21 Bifurcations of Periodic Motions. Period Doubling

In the previous lecture we saw that chaotic motions arise when there are no stable periodic motions in the attracting region of phase space. It is clear that the behavior of any system depends on some control parameters. That is, for some values of the control parameter the system behaves in a very predictable way while for other values the behavior is chaotic. This means that at some value of the parameter there are some stable periodic trajectories. When we change the parameter, some of these trajectories become unstable, and at some critical value of the parameter the last trajectory becomes unstable, and the behavior of the system becomes chaotic. Therefore, it is very important to understand how a periodic trajectory losses its stability.

Visually it is easier to understand what happens to the trajectory if we consider three-dimensional phase space. The equations of motion of a general system can be written as

$$\frac{d\mathbf{x}(t)}{dt} = \mathbf{F}(\mathbf{x}(t), r) \tag{135}$$

where $\mathbf{x}(t)$ is an n-dimensional vector; $n = 3$ for us, and r is the control parameter. Let us study the behavior of the system in the neighborhood of a periodic trajectory $\mathbf{x}(t) = \mathbf{x}_0(t) = \mathbf{x}_0(t + \tau)$. In the neighborhood of this trajectory the behavior of the system can be described by linearized equations. Thus, for the deviations of a trajectory from $\mathbf{x}_0(t)$, $\delta\mathbf{x}(t) = \mathbf{x}(t) - \mathbf{x}_0(t)$, we can write a linear equation with periodic coefficients. The Floquet theorem says that the solution of such an equation can be written in the form $\mathbf{x}(t) = e^{\mathbf{R}t}\mathbf{C}(t)$ where $\mathbf{C}(t)$ is a periodic function with period τ. The stability of the periodic trajectory is determined by the eigenvalues $\gamma_1, \gamma_2, ..., \gamma_n$ of the matrix $e^{\mathbf{R}\tau}$. These numbers are called characteristic multipliers and play a role similar to the roots of the characteristic equation λ_j for two dimensional phase space considered earlier. They describe the amplification, when $|\gamma_j| > 1$, or damping, when $|\gamma_j| < 1$, of small perturbations to the periodic motion along the corresponding eigenvectors. For autonomous systems one multiplier is always equal to 1: there is no stretching or compression in the direction of the phase flow. The stability of the periodic trajectory is determined by the location of the eigenvalues in the complex plane with respect to the unit circle. This is easy to see if we consider the Poincaré section which is shown in Figure 21.1. If $|\gamma_j| < 1$ for all $j \leq n - 1$, the periodic motion is stable. Figure 21.1a illustrates this case when all γ's are real. Then the initial perturbation along the jth direction after each cycle is simply multiplied by a number greater than one in absolute value. Obviously this leads the trajectory away from the periodic orbit. If $|\gamma_j| > 1$ at least for one j the trajectory is unstable Figure 21.1b,c. In the case when some multipliers lie inside the unit circle and some outside, we have a saddle periodic motion Figure 21.1c. If some of the γ's are complex, the trajectory will rotate about the periodic trajectory Figure 21.1d.

This is illustrated in Figure 21.1:

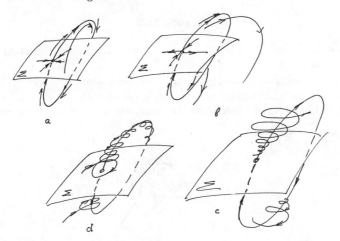

Figure 21.1

In the language of stability multipliers the bifurcation associated with a change of stability of a periodic orbit occurs when one or several multipliers leaves the interior of the unit circle. This can happen in three basic ways.

The first is when one real multiplier passes through the point $\gamma = 1$. When $\gamma_j = 1$, the perturbation along the associated eigenvector does not grow or decay. Therefore any trajectory close to the original one will also be periodic with the same period as the original Figure 21,2. If we reverse the sign of the perturbation, the situation will be exactly the same. The behavior of perturbations orthogonal to the eigenvector whose multiplier passes through the circle does not change much. Therefore, the stability of new periodic trajectories can be determined from a one-dimensional model. For example, when three equilibrium states on a line appear from a stable one, the central state becomes unstable and two new states are stable (Figure 21.2a).

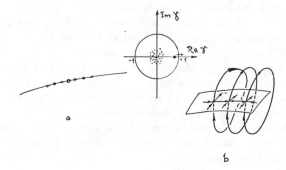

Figure 21.2

So after the bifurcation we will have three periodic orbits: two stable and one unstable Figure 21.2b.

The second possibility is that two complex conjugate multipliers pass through the unit circle. Complex multipliers correspond to rotation about the original periodic trajectory. When the multipliers lie on the unit circle, the perturbations neither grow

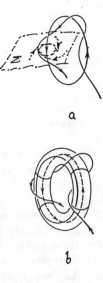

Figure 21.3

nor decay. In the Poincaré section we will see a limit cycle Figure 21.3a. As we know, a limit cycle in the section plane corresponds to quasi-periodic motions. Therefore, as in the first case, a new limit set appears in the neighborhood of the old periodic trajectory, but this time the new limit set is a torus Figure 21.3b.

This is illustrated in Figure 21.3.

The third possibility is that one real multiplier passes through the point $\gamma = -1$. Let us again follow the evolution of a perturbation along the j-th eigenvector when $\gamma_j = -1$. After one period the perturbation is multiplied by -1 Figure 21.4a. After two periods the perturbation is multiplied by $+1$, that is, it returns to the same point (Figure 21.4b)! Therefore, after this bifurcation we will be left with one unstable period 1τ trajectory and one stable trajectory with period 2τ Figure 4c. This is a period doubling bifurcation.

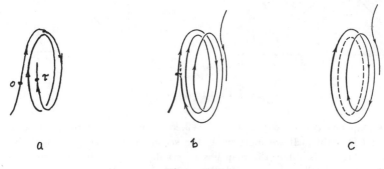

Figure 21.4

All these bifurcations can lead to the onset of chaos in the system. The three routes to chaos are called respectively: intermittency, quasi-periodicity and period doubling.

Let us illustrate how a series of period doubling bifurcations leads to the onset of chaos. To make the discussion clearer, let us consider a one dimensional map. Why can we do this? As we saw before, the behavior along some direction changes qualitatively only when the corresponding multiplier crosses the unit circle. When the multipliers move within the unit circle, we can see only quantitative changes. Therefore the qualitative picture will not change, if we move all but three multipliers to a small region around the origin. The two multipliers which can not be moved without constraint are the multiplier corresponding to the direction tangent to the phase flow and the multipliers responsible for the bifurcation. We saw already in the example of our electric circuit with feedback that, if only two multipliers are not close to zero, we can describe a continuous system using a discrete Poincaré map. Therefore, in general the behavior of a continuous and one dimensional discrete system will be qualitatively the same when we discuss the period doubling bifurcation.

The most studied map to discuss the transition to chaos via period doubling is called the logistic map:

$$x(n+1) = f(x(n)) = 4\lambda x(n)(1 - x(n)) \tag{136}$$

where λ is the control parameter Figure 21.5.

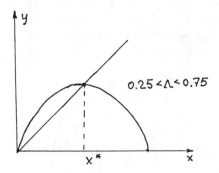
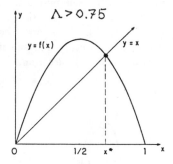

Figure 21.5

This map has two fixed points if $\lambda > 0.25$. The one at zero is unstable, and the second is stable. If we increase λ, the second point becomes unstable when $\lambda = 0.75$. After this value, the map has no stable fixed points. The iterated map $x(n+2) = f(f(x(n)))$ does have stable fixed points Figure 21.6a. This means that the period 1 orbit (a fixed point) becomes unstable and a stable period 2 orbit appears. If we increase the control parameter further these orbits in turn become unstable

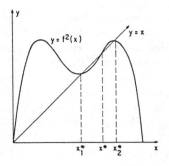
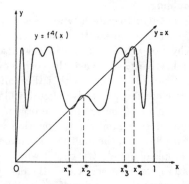

Figure 21.6

but then stable period 4 orbits appear Figure 21.6b, and so on. As the parameter λ increases the bifurcations occur more and more frequently, and finally at $\lambda > 0.892$ there are no stable periodic orbits left and the behavior becomes chaotic. Feigenbaum (1978) noticed that the period doubling sequence converges geometrically:

$$\lambda_\infty - \lambda_n \approx \frac{1}{\delta^n}. \tag{137}$$

In other words

$$\delta = \lim_{n \to \infty} \frac{\lambda_{n+1} - \lambda_n}{\lambda_{n+2} - \lambda_{n+1}}. \tag{138}$$

He found that $\delta = 4.6692016\ldots$. The most important result is that **this number is the same for all period doubling sequences for all maps having one smooth maximum.** Therefore δ is a universal constant, and the behavior of different maps with one maximum is also universal as they pass through this sequence of period doubling bifurcations.

22   Controlled Nonlinear Oscillator. Intermittency

In the previous lecture we discussed the three most important bifurcations of periodic trajectories and said that each of them can lead to chaos. We then studied the transition to chaos through the sequence of period doubling bifurcations using the example of the logistic map. In the logistic map there is only one control parameter and only one route to the onset of chaos. In many physical systems, however, we have more than one control parameter and these systems can approach the chaotic region in many ways. Today and in the next lecture we will consider a system which can become chaotic after a period doubling sequence or after intermittency. This system is the nonlinear oscillator shown in Figure 22.1 coupled parametrically with a control device:

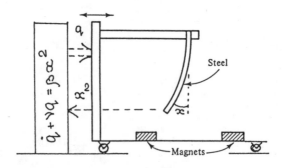

Figure 22.1

The equation of the oscillator without any external force is

$$\ddot{x} + h\dot{x} + (1 - x^2)x = 0 \tag{139}$$

122

There are two attractors, and wherever we start the system, it will evolve towards one of them. We already studied this system when we discussed spontaneous symmetry break down. What will happen if there is an external force? Intuitively it is clear that if the system approaches a hill and has only enough energy to reach the top, any external force will be the decisive factor in determining whether the system will end up in the left attractor or in the right one. It is also clear that if the amplitude of the external force is large enough the system will not stay in either valley but will bounce back and forth; that is, the stable equilibrium points will disappear. The top of the potential hill is the factor which makes the motions unstable and, therefore, we can expect that at some amplitude of the external force the system will behave chaotically.

For reasons which will become clear later, we would like to consider not just an independent parametric force, but a force controlled by motions of the oscillator itself:

$$\ddot{x} + h\dot{x} - (1 - x^2) + qx = 0$$
$$\dot{q} + \nu q = \beta x^2 \tag{140}$$

This is an autonomous system with three dimensional phase space. We can sketch the phase portrait of this system. We know that the system is symmetric with respect to the q-axis; that is, in the $(x, dx/dt)$-plane with $q = 0$ there are two stable foci and one saddle point at the origin Figure 22.2a. Clearly the coordinates of these points satisfy the second equation of our system and, therefore, these points will be equilibrium points in three dimensional phase space too. From the same equation you can see that the saddle is stable with respect to perturbations in the q-direction while the foci are unstable, if $\beta > 0$, with respect to such perturbations. The stable manifold of the saddle is two dimensional, the unstable manifold, one. Therefore the "skillet" of the phase portrait looks as is shown in Figure 22.2b.

It is also easy to see that for $\beta < 0$ the system is globally stable. This is because we can find a Lyapunov functional

$$F = \frac{\dot{x}^2}{2} + \frac{x^4}{2} + \frac{(q-1)x^2}{2} - \frac{a}{4\beta}q^2 \tag{141}$$

The time derivative of the Lyapunov functional is

$$\frac{dF}{dt} = -h\dot{x}^2 + \frac{\dot{q}^2}{2\beta} < 0, \tag{142}$$

when $\beta < 0$. Therefore, as time goes by the Lyapunov functional can only decrease towards its minimum. This means that in the limit of infinite time the system will be in a stable equilibrium state. In our case that state may be one of the foci both of which are stable when $\beta < 0$.

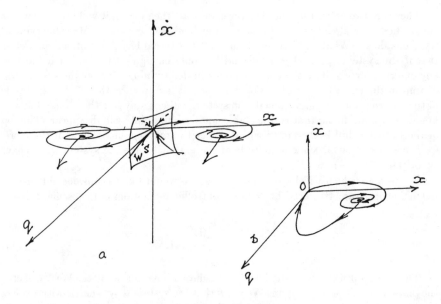

Figure 22.2

What happens if $\beta > 0$? Unfortunately any derivation of an analytical results for this case is based on very difficult mathematics. So we will only adduce the main rigorous results without any proof.

The first thing one can prove is that for $\nu, \beta, h > 0$, there exists a separatrix loop Γ shown in Figure 22.3.

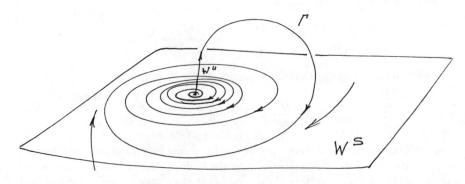

Figure 22.3

The second analytical result is Shilnikov's theorem. This theorem can be applied to any system which can be written in the form $\frac{dx_i}{dt} = f_i(x_1, x_2, x_3)$ $(i = 1, 2, 3)$, if there is a separatrix loop Γ in the phase space of this system. The existence of the separatrix loop Γ implies also that the system has a saddle focus with characteristic exponents λ_i such that $\lambda_1 = \lambda_2^*$ and $\lambda_3 > 0$. The theorem says that if we define α by $\alpha = Re\lambda_{1,2} < 0$, then if $\lambda_3 + \alpha > 0$, the trajectories in the neighborhood of Γ form a so-called homoclinic structure which contains a countable set of saddle periodic trajectories. In this case, as we know from our previous lectures, the behavior of the system becomes chaotic: a phase trajectory wanders "randomly" between the unstable periodic trajectories. This theorem can be applied to our system, and, therefore, proves our intuitive guess that it can demonstrate chaotic behavior.

Now let us rewrite our system in a more standard form. For this purpose let us change variables and parameters as follows

$$x = \frac{\epsilon}{\sqrt{2\sigma}}X; \quad \dot{x} = \frac{\epsilon^2}{\sqrt{2}}(Y - X);$$

$$q = \epsilon^2(Z - \frac{X^2}{2\sigma}); \quad t_{old} = \frac{\sqrt{\sigma}}{\epsilon}t$$

$$h = \epsilon\frac{\sigma + 1}{\sqrt{\sigma}}; \quad \nu = \epsilon\frac{b}{\sqrt{\sigma}};$$

$$\beta = \epsilon\frac{2\sigma - b}{\sqrt{\sigma}}; \epsilon = \sqrt{r - 1} \tag{143}$$

$$\dot{X} = -\sigma(X - Y)$$
$$\dot{Y} = rX - Y - XZ$$
$$\dot{Z} = -bZ + XY \tag{144}$$

and our equations turn into the Lorenz system which was first derived for the approximate description of thermoconvection and has become a 'canonical' model of low dimensional chaos.

The phase portrait of this system for standard values of the parameters ($\sigma = 10, b = 8/3, r = 28$) is seen in Figure 22.4 while the chaotic time series for $X(t)$ is shown in Figure 22.5.

In the phase space of the Lorenz system there is an attracting limit set whose trajectories are all unstable. This limit set is called a chaotic attractor. The trajectories on this attractor are very close to a two dimensional surface. To prove this statement let us consider the volume $V(t)$ of a drop of phase liquid (a set of initial conditions) as a function of time. We can write $\frac{dV}{dt} = \nabla \cdot \mathbf{F}$, where $\mathbf{F}(\mathbf{x})$ is defined by $\frac{d\mathbf{X}}{dt} = \mathbf{F}(\mathbf{x})$. Therefore, for the rate of change of the volume we have

$$\frac{dV}{dt} = \frac{\partial \dot{x}}{\partial x} + \frac{\partial \dot{y}}{\partial y} + \frac{\partial \dot{z}}{\partial z} = -(1 + \sigma + b), \tag{145}$$

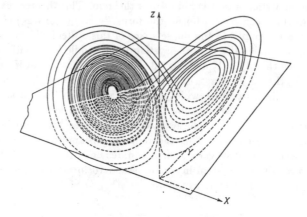

Figure 22.4

Figure 22.5

which is negative and large in magnitude for the standard choice of parameters. The phase drop becomes flat very fast, and as time approaches infinity the drop becomes almost two dimensional. This means that in this case we again can use the Poincaré section method.

We choose the secant surface so that it passes through all $Z(t)$-maxima of the trajectory for one "wing" Figure 22.6 although the choice of the surface is not unique.

For the standard parameter values the map constructed this way is shown in Figure 22.7. This map looks like the one we used to describe the motions in the nonlinear electric circuit. We again have a region corresponding to the unwinding spiral (Part A) and a region returning the trajectory back to the neighborhood of the unstable focus (Part B). But the behavior in this system depends on the control parameter r. At some r the system behaves chaotically, at other values of r it shows very predictable behavior. Then the question arises about the way the system changes its behavior as the control parameter is varied.

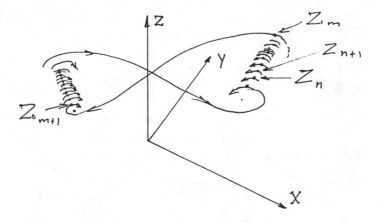

Figure 22.6

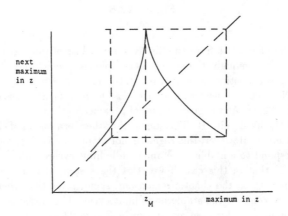

Figure 22.7

We already know one way to chaos; namely, a sequence of period doubling bifurcations. Can this be the way to chaos in Lorenz system? The answer is yes. Look at Figure 22.8 where the maps for different values of r are shown. You can see that there is no chaos in the map when $r > 30.2(\sigma = 10, b = 8/3)$ Figure 22.8a. The fixed point of this map corresponds to a periodic trajectory of the original system. As we decrease r the slope of the mapping function at the fixed point becomes larger, and after it reaches -1, the fixed point becomes unstable. Then a stable limit cycle with doubled period is formed Figure 22.8b. After a sequence of such bifurcations the Lorenz chaotic attractor is formed (Figure 22.8c).

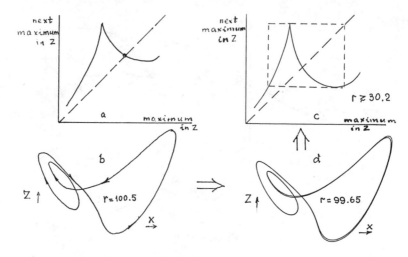

Figure 22.8

All this is familiar to us from the discussion of orbits of the logistic map. When parameter r is large enough, the system is completely predictable. As we decrease r we observe the transition to a chaotic state via a sequence of period doubling bifurcations. Now let us consider another case. The map for the case of $r < 24.06$ is shown in Figure 22.9a. At the given values of parameters there are no stable periodic motions about the fixed point B . This means that there is a region of chaotic motions near this fixed point. However, this region is not attractive at this value of r. Points C and A correspond to a stable node and saddle limit cycles with the same period. It is easy to see that in this case if we start the system at some arbitrary initial condition, it may go to the chaotic region and spend some time there. After that it will eventually come to the stable node limit cycle. The trajectory and the time series corresponding to this case are shown in Figure 22.9a. As we increase r the two limit cycles A and C are getting closer, and at $r = 24.06$ they annihilate Figure 22.9b. This is a local bifurcation, and hence the trajectory coming from outside does not know that the stable cycle has disappeared. It still goes to the region where the stable limit cycle existed, and bounces in this region in search of the limit cycle. After a while the trajectory realizes that there is no a stable state in that region and leaves it for the chaotic region. The map for $r > 24.06$ is shown in Figure 22.9c. It is easy to see that if $r \simeq 24.06$, the system spends almost all its time in the search for the missing stable periodic trajectory. The amplitude of the oscillations during this period changes very slowly. But then suddenly the system goes to the region of chaotic motions and makes a few bursts. After that it returns to check the region where the stable periodic trajectory was and repeats the whole thing again. The

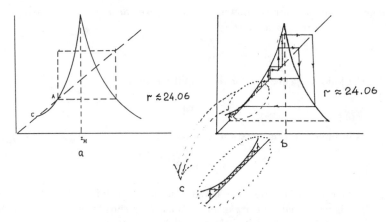

Figure 22.9

time series therefore will look like a laminar motion interrupted with turbulent bursts Figure 22.9c.

The general scenario of transition to chaos we just described is called transition to chaos through intermittency.

It is shown in Figure 22.10.

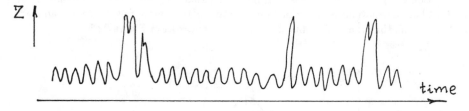

Figure 22.10

Can we describe the behavior of the system close to the bifurcation point analytically? Of course, since the motion is chaotic, we can not integrate the equations of motion. But we can find some averaged description of the motions. For instance, let us find the average time the system spends in the laminar region. To do this let us approximate the mapping function in the laminar region by a quadratic polynomial: $Z_{n+1} = f_r(Z_n) = r^* + Z_n + aZ_n^2$ where $a > 0$ and r^* is the difference between r and $r_{crit} = 24.06$. By scaling we can choose $a = 1$. If the system is very close to the bifurcation point, the tunnel of laminar motions is very narrow, and the amplitude

Z_n changes very little after each step. Therefore, we can make the transition from discrete time to continuous time:

$$Z_{n+1} - Z_n = \frac{\Delta Z}{\Delta n}\bigg|_{\Delta n=1} = r^* + Z_n^2,$$ (146)

and write the following equation for the amplitude $Z(t)$:

$$\frac{dZ}{dt} = r^* + Z^2$$ (147)

One can easily integrate this equation to obtain

$$t\sqrt{r^*} = \arctan\left(\frac{Z(t) - Z^*}{\sqrt{r^*}}\right),$$ (148)

where Z^* is the center of the region of laminar motions which is visited by the system at $t = 0$. If r^* is very small, the argument of the arctan runs from $-\infty$ to $+\infty$, while the system passes from one end of the tunnel to the other. The average time the system spends in this region is given by

$$<t> \sim \frac{\pi}{\sqrt{r^*}}$$ (149)

The average time is defined by

$$<t> = \int_0^\infty p(t, r^*)\, t\, dt \sim \frac{1}{\sqrt{r^*}}$$ (150)

where $p(t, r^*)$ is the probability density for the observation of a laminar stage with the duration t at some given r^*. It is clear intuitively that if $r^* \simeq 0$, then the probability of short laminar motions is very small because the motion on average is almost laminar. This means that the probability density must appear as in Figure 22.11.

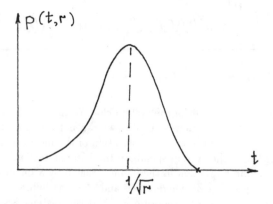

Figure 22.11

Let us summarize the main results of these two lectures. First of all, we learned that the period doubling bifurcations sequence is indeed a general way to chaos. Secondly, we discovered a new way to chaos: intermittency, which is the result of the annihilation of node and saddle limit cycles. And finally, we found that there may be more than one way to chaos in the same dynamical system.

23 Scenarios of the Onset of Chaos. Chaos through Quasi-Periodicity

When we discussed bifurcations of periodic motions, we said that each of them correspond to a specific scenario of arising of chaos in the system. The first scenario we considered was the sequence of period doubling bifurcations. In spectral language we start with the spectrum shown in Figure 23.1a; T_0 is the period of original motion. After the first bifurcation the base frequency becomes half of the original, and we observe the spectrum shown in Figure 23.1b. Similarly after the second bifurcation, we will see a spectrum with the base frequency one quarter of the original, Figure 23.1c. At the critical value of the control parameter the spectral lines produced during the bifurcation sequence merge and form a continuous spectrum corresponding to chaotic motion in the system Figure 23.1d:

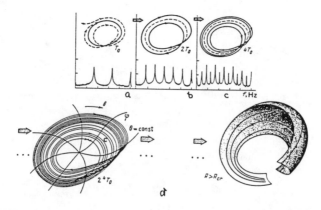

Figure 23.1

The second scenario which we studied was the transition to chaos through intermittency. In this case we again start with the spectrum of a periodic signal Figure 23.2a. As we change the control parameter the time series changes from being completely periodic to the one shown in Figure 23.2b. The spectrum of this signal is close to that of a periodic signal, but the spikes will be wider Figure 23.2c. As we change the control parameter further the average duration of periodic motion decreases, and the spectrum becomes wider Figure 23.2d; the behavior becomes more and more chaotic.

Intermittent transition to chaos:

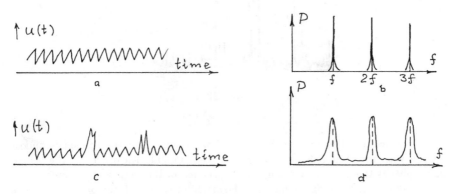

Figure 23.2

The last bifurcation sequence we wish to discuss is creation of a torus in the phase space which corresponds to quasi-periodic motions. We understand quite well how the spectrum changes after the first bifurcation. We start with the same spectrum as in the previous cases Figure 23.3a. After the bifurcation the spectrum is that of quasi-periodic motion Figure 23.3b. What happens after this point? How can chaotic motion appear in this system? Let us find out.

Right after the bifurcation point the evolution of the system is that of two modes with incommensurate frequencies. The further development of the system as we change the control parameter will be determined by the interaction of these modes and their harmonics. We have already studied one possibility when we discussed the synchronization phenomenon. If the coupling between the modes is strong enough, we can observe mode locking. When the modes are locked the behavior of the system changes from quasi-periodic to periodic. We can study this phenomenon using the averaging method as we did before. But as you remember the averaging method, as any approximate method, can be applied only in a very limited region of parameters.

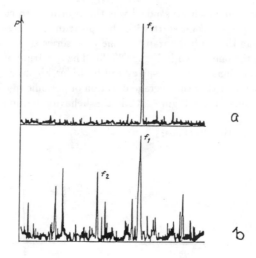

Figure 23.3

More specifically, when we use this method we assume that the motion is almost periodic with amplitudes changing very little over a period. The other way the system may evolve as the parameter changes is destruction of the torus and emergence of a strange attractor. Since at the moment the attractor is formed, the periodic motions cease to exist this transition can not be described in the framework of averaged equations.

Not to be solely abstract let us consider our favorite system, that is a pendulum with external periodic forcing.

$$I\frac{d^2\theta}{dt^2} + \frac{d\theta}{dt} + \beta\sin\theta = F + Q\cos\omega t \tag{151}$$

The external force looks a little unusual. $Q\cos\Omega t$ is the ordinary periodic torque. But F is a constant torque which does not depend on the position of the pendulum. Let us assume that the dissipation is so high and that the frequency of the external force is so low that we can neglect the inertia of the pendulum: $I\frac{d^2\theta}{dt^2} \simeq 0$. Our equation then becomes

$$\frac{d\theta}{dt} + \beta\sin\theta = F + Q\cos\Omega t, \tag{152}$$

and this proves a convenient framework for the investigation of synchronization and quasi-periodic motions. In this equation we have two characteristic periods: one is associated with the action of F, the other is the period of the external harmonic force. We will consider the case when β is small. Then we can look for the solution in the form of an asymptotic series:$\theta(t) = \theta^{(0)}(t) + \beta\theta^{(1)}(t) + \ldots$. We can find $\theta^{(0)}$ and $\theta^{(1)}$ assuming that at time t', θ equals $\theta(t')$.

$$\theta^{(0)}(t) = \theta(t') + F(t - t') + \frac{Q}{\Omega}(\sin\Omega t - \sin\Omega t')$$

$$\theta^{(1)}(t) = -\int_{t'}^{t} \sin\theta^{(0)}(\tau)d\tau. \tag{153}$$

If now we let $\theta_n = \theta(t_n)$, where $t_n = \frac{2\pi n}{\Omega}$, we will reduce equations (153) to the Arnol'd map which is the classical system for discussion of transition to chaos via quasi-periodicity. This map is also known as the circle map:

$$\theta_{n+1} = \theta_n + \frac{2\pi F}{\Omega} + K(\frac{2\pi F}{\Omega}; \frac{Q}{\Omega})\sin(\theta_n + \phi + \pi). \tag{154}$$

This map can be also written in the form

$$\theta_{n+1} = \theta_n + w + K\sin\theta_n. \tag{155}$$

In this map the nonlinear term $K\sin\theta_n$ models the effect of nonlinear coupling between the two periodic motions: the one due to the constant torque, and the one due to the periodic forcing.

In the next lecture we will discuss various properties of motions which can help us to distinguish between simple, regular motion and chaotic ones, and also allow us to determine how "strong" chaos is in a given system. But there is one parameter which is used specifically to study the synchronization and transition to chaos via quasi-periodicity. We now introduce this parameter called **winding number**. The winding number can be thought of as of the average increase of the angle θ per step or as of the average frequency:

$$W = \lim_{n \to \infty} \frac{f^n(\theta) - \theta}{2\pi n}, \tag{156}$$

where $f(\theta) = \theta + w + K\sin(\theta)$ in our case. $f^n(\theta)$ is the result of applying the nonlinear function f n times. In the mode locked state, the motion is periodic, $\theta_{n+q} = \theta_n + 2\pi p$, and $W = p/q$ is a rational number. The winding number is irrational for a quasi-periodic motion and is not defined for chaotic motion.

For $K = 0$ clearly $W = w$, and we will have quasi-periodic motion unless w is a rational number. But rational numbers have zero measure on the w-axis, and, therefore, we are most likely to see quasiperiodic motions. As K increases, the regions where W is rational are no longer points but some intervals around rational numbers. At $K = 1$ the measure of periodic motions equals one while that of quasi-periodic motion is zero. If we plot the winding number as a function of w we will see the object which is known as the devil's staircase Figure 23.4.

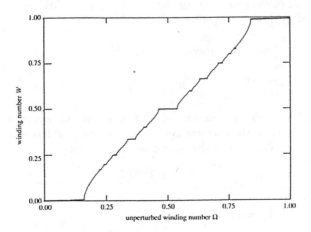

Figure 23.4

The intervals where the graph is horizontal are the synchronization intervals with a certain rational winding numbers.

We can generalize this diagram for the case of arbitrary K Figure 23.5. The regions of synchronization which start from rational numbers on the w-axis are called Arnol'd tongues.

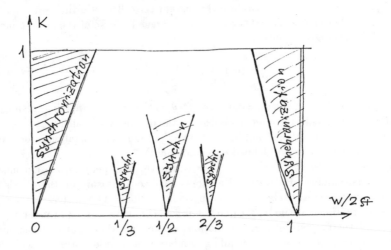

Figure 23.5

You can see that a section of this graph with the line $K = 1$ is a manifold with a noninteger dimension (this term is to be discussed in the next lecture) which is called a Cantor set. If we consider this section and take into account the winding numbers corresponding to each tongue we will obtain the devil's staircase.

When the system is in the synchronization region with two frequencies, we observe periodic pulsation which corresponds to a closed trajectory on a two dimensional torus. As K becomes larger than 1 the description of synchronization based on perturbation methods will fail, the Arnol'd tongues will start merging, the periodic trajectory corresponding to the synchronization regime will either disappear or change its stability, and the motion will become chaotic. At this moment the two torus disappears creating a strange attractor.

The spectrum suddenly broadens as indicated in Figure 23.6.

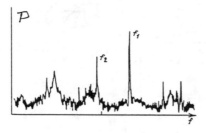

Figure 23.6

Now we can collect everything we know about the transitions to chaos in a simple table

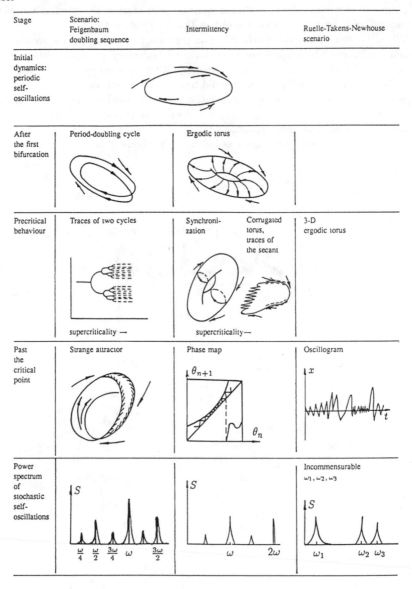

Stage	Scenario: Feigenbaum doubling sequence	Intermittency		Ruelle-Takens-Newhouse scenario
Initial dynamics: periodic self-oscillations				
After the first bifurcation	Period-doubling cycle	Ergodic torus		
Precritical behaviour	Traces of two cycles	Synchronization	Corrugated torus, traces of the secant	3-D ergodic torus
	supercriticality →	supercriticality→		
Past the critical point	Strange attractor	Phase map θ_{n+1} ... θ_n		Oscillogram x ... t
Power spectrum of stochastic self-oscillations	S ... $\frac{\omega}{4}$ $\frac{\omega}{2}$ $\frac{3\omega}{4}$ ω $\frac{3\omega}{2}$	S ... ω 2ω		Incommensurable $\omega_1, \omega_2, \omega_3$ S ... ω_1 ω_2 ω_3

24   Characteristics of Chaos.
Experimental Observation of Chaos

So far we have discussed only various theoretical aspects of chaos. We learned that
if we have complete knowledge of the nonlinear system, we can predict the kind of
behavior corresponding to the given values of its parameters. However, if we do not
know the equations describing the system and try to get some information about it
from an experiment everything becomes more complicated, and from the physicist's
point of view much more interesting. One problem is that any measurement contains
some level of contamination. We should learn to differentiate between noise and a
chaotic dynamical signal, since to the linear observer, they look much the same. It
turns out that neither intuition nor standard linear methods help us with this. Each
of the time series shown in Figure 24.1 has a wide band Fourier spectrum and rapidly
decaying autocorrelation function, but one of them was generated by a dynamical
system while another one comes from a random number generator and is classified as
"noise".

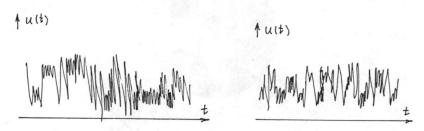

Figure 24.1

Is there a way to distinguish these two signals? The answer is "yes". This is
because "noise" is generated by a system with many, many degrees of freedom while

a chaotic signal is produced by a nonlinear system with a few degrees of freedom, and, therefore, we expect that there are some characteristics of the chaotic signal which are very different from those of the "noise". `Lyapunov exponents, dimensions and entropy` are some of these quantities.

24.1 Fractal Dimension

We already noticed that the dimension of a chaotic limit set is not always an integer. We can generalize the common definition of dimension for these sets. Let us cover the attractor with cubes (or spheres) of size ϵ, and let $N(\epsilon)$ be the minimum number of such cubes required to cover the attractor completely. We define the dimension D_A of the attractor by the limit

$$D_A = \lim_{\epsilon \to 0} \frac{\log N(\epsilon)}{-\log \epsilon}. \tag{157}$$

Let us make sure that this definition works well for all known cases. For a D dimensional set of volume V the number of cubes we need to cover the set completely is given by $N(\epsilon) \propto \frac{V}{\epsilon^D}$ for small ϵ. You can easily prove that for a finite set of points the D following from this definition is zero, for a line, it is unity, for an area, it is 2, and for a three dimensional figure, it is 3. To see how this definition works for more complicated sets let us consider the set shown in Figure 24.2.

Figure 24.2

One can see that tripling the radius of a sphere leads to five times as many elements within the sphere. The number of elements grows faster than in the one dimensional case but slower than in two dimensions. This is because the set is very porous or holey and almost all of two space is empty. Indeed, the dimension of this set is noninteger and equals $D_A = \frac{\ln 5}{\ln 3} \approx 1.46$.

We know that attractors appear in dissipative systems, and, therefore, the dimension of a strange attractor is always smaller than that of the number of degrees of freedom of the system itself. In other words phase volume always decreases in a dissipative system, and since the system approaches the attractor as time goes to infinity, the volume of attractor in the original phase space is zero. However, we can extend the definition of volume for the case of fractal sets and assume that a cube of size ϵ in D dimensions is ϵ^D. It is clear that we defined the dimension D in such a way that the volume of a strange attractor in D space is nonzero and is given by the formula

$$V_D = \lim_{\epsilon \to 0} N(\epsilon)\epsilon^D. \tag{158}$$

Dimensions are entirely geometrical characteristics. However, we saw that they is somehow connected with the evolution of the phase volume. Therefore, we expect being able to estimate the dimension using the characteristics of phase volume evolution.

24.2 Lyapunov Exponents

Let us consider a trajectory $\mathbf{X} = \mathbf{X}_0(t)$ in the phase space of the system $\frac{d\mathbf{X}}{dt} = \mathbf{F}(\mathbf{X}, \mu)$, and the deformation of the hypersphere of radius $|\mathbf{X}_0(t = 0)|$ as the trajectory evolves. If both the sphere and its deformations during the evolution are small, we can linearize the equations of motion to find the evolution equation for $\xi(t) = \mathbf{X}(t) - \mathbf{X}_0(t)$, where $|\xi(0)| \ll 1$:

$$\frac{d\xi_i(t)}{dt} = \frac{\partial F_i(\mathbf{X}_0(t), \mu)}{\partial X_k}\xi_k(t). \tag{159}$$

As the trajectory evolves, the volume element contracts in some directions and stretches in others and the hypersphere transforms into an ellipsoid (Figure 24.3). During the motion the directions of the semiaxes of the ellipsoid change as well as their lengths. We call the lengths $l_j(t)$, where j labels the eigendirections of the Jacobian matrix. The limits

$$\lambda_j = \lim_{t \to \infty} \left(\frac{1}{t}\right)\left(\frac{\log l_j(t)}{\log |\xi(0)|}\right) \tag{160}$$

are referred to as the Lyapunov characteristic exponents. These numbers are very important for experiment, because, if at least one Lyapunov exponent is positive, and the observed signal remains bounded, one can be sure that she is dealing with a chaotic signal.

The deformation of a sphere of initial conditions:

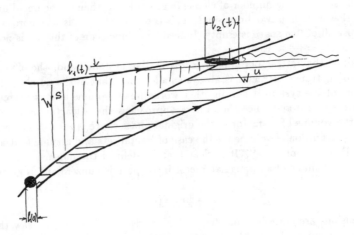

Figure 24.3

The sum of Lyapunov exponents determines the average along the trajectory of a volume in the phase space. The local change of the volume at each point along the trajectory is specified by

$$\nabla \cdot \frac{d\mathbf{X}}{dt} = \nabla \cdot \frac{d\xi}{dt} = \text{Trace}(\frac{\partial \mathbf{F}(\mathbf{X}, \mu)}{\partial \mathbf{X}}). \tag{161}$$

One can prove that the average value of this divergence along the trajectory in a system with N degrees of freedom is given by

$$\lim_{t \to \infty} \frac{1}{t} \int_0^t \nabla \cdot \dot{\xi}(t') dt' = \sum_{j=1}^{N} \lambda_j. \tag{162}$$

For a dissipative system this sum is always negative, i.e. arbitrary volumes contract in N dimensional phase space. Let us arrange the Lyapunov exponents in the order $\lambda_1 \geq \lambda_2 \ldots \geq \lambda_N$. Now if we can make the sum

$$\sum_{j=1}^{m} \lambda_j \tag{163}$$

become zero at some m, this would mean that the volume in some subspace of dimension m is constant, and we could say that there is an m dimensional attractor in the phase space of our system. Let M be the largest integer such that the sum is still positive when $m = M$ but becomes negative if m = M + 1. So the volume shrinks in M+1 dimensional phase space and is not defined in M dimensional phase space. To make the volume constant in some subspace we should take into account a fraction of the contracting action of λ_{M+1}. We can define a so-called **Lyapunov dimension** by

$$D_L = M + \frac{\sum_{j=1}^{M} \lambda_j}{-\lambda_{M+1}} = M + \frac{\sum_{j=1}^{M} \lambda_j}{|\lambda_{M+1}|}. \tag{164}$$

Since we neglected all the other negative exponents, this Lyapunov dimension is, in general, an upper limit for the fractal dimension defined by counting boxes: $D_L \geq D_A$.

The sum of positive Lyapunov exponents is called the Kolmogorov-Sinai entropy and characterizes the rate of divergence of phase trajectories.

You could now say: "All this is very good. But we gave these definitions as if we knew all the dynamical variables. In a real experiment we usually do not have this much luck and measure just one variable or possibly a few physical quantities." This is indeed correct, but it turns out that the measurement of just one dynamical variable is enough to learn a lot about the structure of the underlying chaotic limit set. The idea of treating this kind of real data was proposed by Ruelle. The idea is to determine the characteristics not in the original phase space of true dynamical variables but in the *reconstructed phase space* of d dimensional vectors:

$$\mathbf{x}(j) = [u(j), u(j+1), \ldots, u(j+d-1)] \tag{165}$$

where $u(j) = u(t_0 + j\Delta t)$ is the measured quantity, and Δt is the time interval between observations. See Figure 24.4a. If we follow the evolution of this vector as j increases, we will be able to define a trajectory in this new phase space. Now we can, in principle, compute dimension and Lyapunov exponents for the limit set in this new space, and one can prove that these (and many other) characteristics of the reconstructed limit set are the **same** as those of the true chaotic attractor in the true phase space of the true dynamical variables.

Now, let us speculate a little about how we can differentiate between pure noise and a dynamical signal using this method. You, probably, noticed that we did not say anything about how to choose the dimension of the reconstructed space d. Suppose we are working with a dynamical signal. In this case we need some limited amount of information to predict the state of the system at the next time step. Let us start with 'time' n = 1 by which we mean $t = t_0 + \Delta t$. One measurement does not give enough information, in general, to predict the next state. However, if we increase the number of pieces of information and consider d = 2 in making up the data vector, the characteristics of the limit set may change significantly. If we keep increasing d, we will reach a value of d for which d independent measurements of the observable suffice to predict the next state of the system in time. Hence, further increases of d will not cause meaningful changes of the characteristics of the limit set. See Figure 24.4b. We would then know the correct d, which is called the **embedding dimension**, and would be able to evaluate the correct values of the fractal dimension etc. However if we are dealing with `noise`. it is clear that we can not predict the value of an observable at the next time step regardless how well we know the history of the signal. This means that in this case all characteristics of the 'attractor' will seem to change as we increase d without any saturation in their values. This is the feature of noise that allows us to distinguish it from a dynamical signal.

So it seems as if we have quite clear plan of action. However, it turns out that the algorithms for the computation of dimension and Lyapunov exponents are often very time consuming, poorly convergent or not precise enough. Because of this some physicists prefer to compute a different characteristic of time series called correlation integral. This characteristic is defined by

$$C_d(\epsilon) = \lim_{N \to \infty} \frac{\text{Number of pairs of points with distance between them less than } \epsilon}{N(N-1)}$$

$$= \lim_{N \to \infty} \frac{1}{N} \sum_{i=1}^{N} [\frac{1}{N-1} \sum_{j=1; j \neq i}^{N} \Theta(|\mathbf{x}(i) - \mathbf{x}(j)| - \epsilon)], \qquad (166)$$

where $\mathbf{x}(i)$ is the d dimensional vector made out of time delayed versions of the observable $u(n)$ and $\Theta(\bullet)$ is the Heaviside function.

The quantity

$$d_C = \lim_{\epsilon \to 0} [\frac{\log C_d(\epsilon)}{\log \epsilon}], \qquad (167)$$

is called the correlation dimension. This dimension turns out to be very good estimate for the box counting dimension introduced before.

Note three things. First, the correlation integral is a function of the dimension d used to form the vectors. Second, the correlation function can be defined for any time series (chaotic, noisy and even mixed) without referring to the idea of limit set reconstruction. Finally, the correlation integral is a function of ϵ, and this allows us to explore the structure of our signal at different scales is phase space.

In reality, of course, we can not make the number of data points N arbitrarily large and ϵ arbitrarily small. Figure 24.4c presents a family of plots of $\log C_d(\epsilon)$ versus $\log \epsilon$ of the kind we are likely to see when we compute the correlation integral from experimental data. If ϵ is very large (that is, as large as the attractor itself) all points will be inside the spheres of radius ϵ, so if we increase ϵ the correlation integral will not change and we will see the horizontal part of the plot for large ϵ. If ϵ is so small that there is only one point in each sphere, and if we decrease ϵ, the correlation integral will not change until the spheres are so small that no data points are in them. We will see a horizontal slope for ϵ so small as well. Then we can see an interval which may be linear in this plot and whose slope approaches a constant as d increases. This is the part which gives us information about the dynamical component of the signal. The slope of the plot is the correlation dimension, d_C.

Finally, there is a part of the plot which seems to become steeper and steeper as we increase d implying that the embedding and correlation dimensions are both infinite. This part of the plot corresponds to the noise which is always present in any measurement.

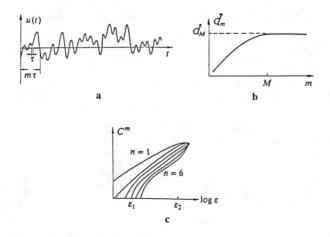

Figure 24.4

You see that one cannot define the correlation dimension as the slope of the plot in Figure 24.4 at small ϵ. Instead one should consider the local slope of the plot. This analysis can help to understand the structure of the signal. It may indicate how many additive dynamical components are contributing to the observed signal, what their dimensions are and what the level of noise is. To see how one can obtain this information more clearly let us consider the correlation dimension in the dynamic and in the noisy regions of Figure 24.4 as functions of d (Figure 24.5). You see that the dimension for the dynamical region is saturated at some embedding dimension while the dimension of the noisy component of the signal grows without bound as d increases.

The important thing to understand is that we can determine only the dimension of the limit set of the system. The correlation dimension of the limit set gives a lower limit for the dimension of the system. In fact an infinite dimensional dynamical system can have a low dimensional attractor. But this is the subject of the next lecture.

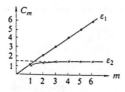

Figure 24.5

See the Table for a summary of these qualitative classes of behavior.

Table   From the *left* to the *right* the respective dimensions D, entropies H, signals $U(t)$ and types of attractors are shown.

Dimension	Entropy	Signal $U(t)$	Type of attractor	
$D = 0$	$H = 0$	$u(t) = $ const	Stable equilibrium	
$D = 1$	$H = 0$		Limit cycle periodic self-excited oscillations	
$D = 2$	$H = 0$		Open winding motion on a 2-d torus. multi-periodic self-excited oscillations	
$D = 2.06$	$H > 0$		Strange attractor (Lorenz attractor) Stochastic self-excited oscillations	
$D \gg 1$ $D \to n$ $n \to \infty$	$H > 0$		Multi-dimensional attractor	

25 Multidimensional Chaos. Discrete Ginzburg–Landau Model

In this last lecture we will consider a more complicated system that exhibits chaotic behavior. When we discussed the bifurcations leading to the appearance of strange attractors we studied only low dimensional systems which could be described by systems of ordinary differential equations. However, there are many systems which can be described only by partial differential equations. and which can also behave chaotically. Indeed, it may fairly be said that the description of any physical process by ordinary differential equations represents a "lumped" parameter or long wavelength approximation to the dynamics of the underlying fields. In this lecture we will consider an example of such continuous systems called the Ginzburg–Landau model, and for simplicity we will restrict ourselves to one spatial dimension. This model is used to describe motions in a variety of excitable media such as chains of Van der Pol generators, a chain of Taylor–Couette vortices, convection and many other systems. But let us consider for a change a biological example of such a medium to illustrate that nonlinear physics is indeed an interdisciplinary science, and also to illustrate how nonlinear phenomena affect our life.

Let us consider the propagation of excitations in a neural membrane. The equations of evolution for one element of the membrane is described by Fitz–Hugh–Nagumo equations:

$$\epsilon \frac{dx}{dt} = x - y - \frac{x^3}{3}$$
$$\frac{dy}{dt} = x - qy. \tag{168}$$

Here x is the "excitation" or "propagator" variable, and y is a "recovery" or "controller" variable. When $\epsilon \ll 1$, the variables x and y oscillate in a relaxation fashion: that is, phase trajectories have fast and slow parts on a stable limit cycle. When

147

$\epsilon = 1$, the limit cycle is close to a circle. We can change the variables to write this system in a more familiar form:

$$\ddot{x} - \mu(1 - \rho x^2)\dot{x} + \mu x - \beta x^3 = 0, \qquad (169)$$

where $\mu = q - 1$, $\rho = (q-1)^{-1}$ and $\beta = \frac{(q-2)}{3}$. This is the Van der Pol–Duffing equation. If μ is very small, the oscillation amplitude will be a very slow function of time, and we can use the averaging method to find the equation for the slow amplitude. The only difference from all our previous examples where we used this method is that here there is a frequency correction due to the nonlinear term. We can write the solution in the form

$$x(t) = a(t)e^{i\Omega(|a|^2)t} + \text{cc}, \qquad (170)$$

where a(t) is the slow amplitude for which we can write a Landau–Stuart equation:

$$\frac{da}{dt} = a[1 - (1 + i\beta)|a|^2]. \qquad (171)$$

This is the equation describing the evolution of each fragment of a neural membrane when they do not interact with each other. Let us introduce a linear coupling among the neighboring membrane elements and assume that it is isotropic. We will label the excitation amplitude by an index j. Then the difference-differential equation describing whole membrane is

$$\frac{da_j}{dt} = a_j[1 - (1 + i\beta)|a_j|^2] + \kappa(1 - ic)[a_{j+1} - 2a_j + a_{j-1}], \qquad (172)$$

and the last term is responsible for the interaction among pieces of the membrane. This is a discrete Ginzburg–Landau equation. The continuous analog of this equation is

$$\frac{\partial a(x,t)}{\partial t} = a(x,t)[1 - (1 + i\beta)|a(x,t)|^2] + \kappa(1 - ic)\nabla^2 a(x,t). \qquad (173)$$

Let us study the difference-differential equation. Before we proceed with analysis we should specify the boundary conditions. For simplicity we will study the case of periodic boundary conditions: $a_j(t) = a_{j+N}(t)$ at any j. This corresponds to a neural membrane of N elements bent into a ring. Let us try to find some particular solutions. The simplest kind of solutions we can think about is a sinusoidal propagating wave:

$$a_j^n(t) = A_n e^{i(\omega_n t + j\theta_n)}. \qquad (174)$$

Here n labels propagating waves with different frequencies. Substitution of this solution into our differential-difference equations yields the nonlinear dispersion law

$$|A_n|^2 = 1 - 4\kappa \sin^2\left(\frac{\theta_n}{2}\right)$$

$$\omega_n = -\beta + 4\kappa(\beta + c)\sin^2\left(\frac{\theta_n}{2}\right), \qquad (175)$$

where $\theta_n = \pm\frac{2\pi n}{N}$, n = 0,1,2,...N/2. For n = 0 we have a spatially uniform solution. For $n = \frac{N}{2}$ ($\theta_1 = \pi$) we have a solution in the form of so called π-oscillations ,when the excitations in any two neighboring fragments of the membrane have the same amplitudes and opposite phases. All other solutions are travelling waves.

We notice that at $\kappa > \kappa^*$ (given in a moment), the only kind of solution which can exists is a homogeneous oscillation, because for all other waves the expression for $|A_n|^2$ gives a complex value of the intensity.

$$(\kappa^*)^{-1} = 4\sin^2(\frac{\pi}{N}) \tag{176}$$

Therefore, we can conclude that at large coupling coefficient the oscillations of all elements are synchronized. As we decrease the coupling the synchronization weakens and shorter waves appear. The smaller the coupling the higher the number of excited harmonics. Each of these harmonics is a periodic motion whose image in the phase space is a limit cycle. Let us study the stability of these trajectories.

First of all, we notice that the system is invariant under the transformation $a_{j\text{new}} = a_{j\text{old}}e^{i\phi}$. Using this we can change variables:

$$b_j(t) = a_j(t)e^{-i(\omega_n t - j\theta_n)}, \tag{177}$$

and rewrite the Ginzburg–Landau equation:

$$\frac{db_j(t)}{dt} = (1 - \omega_n)b_j - (1 + i\beta)|b_j|^2 b_j + \kappa(1 - ic)(b_{j+1}e^{i\theta_n} + b_{j-1}e^{-i\theta_n} - 2b_j). \tag{178}$$

This change of variables brings us into the frame of the traveling waves, and the limit cycles associated with them become fixed points. Now we can consider the stability of periodic motions by linearizing this equation near the solution b_j^0 which is determined by $|b_j^0|^2 = A_n^2$.

Examination of the linearized version of this equation leads to the interesting result that stability of homogeneous oscillations with respect to the lth mode is determined by the Lyapunov exponent

$$\lambda_l = -4 - 4\kappa\sin^2(\frac{\theta_l}{2}) \pm [1 - (4\kappa\sin^2(\frac{\theta_l}{2})^2 + 8\beta c\kappa\sin^2(\frac{\theta_l}{2})]^{1/2}. \tag{179}$$

We can see that when

$$\kappa > \frac{\beta c - 1}{2(1 + c^2\sin^2(\frac{\pi}{N})}, \tag{180}$$

the characteristic exponent λ_l is negative, and homogeneous oscillations are stable. If we want to consider the limit of a continuous medium we should let N go to infinity, then we arrive at the condition for stability of a homogeneous solution for a continuous system: $\beta c < 1$.

The Lyapunov exponents for the trivial equilibrium $a = 0$ are given by

$$\lambda_l = 1 - 4\kappa(1 - ic)\sin^2(\frac{\theta_l}{2}). \tag{181}$$

We see that $\mathrm{Re}\,\lambda_l = |A_l|^2$. Now we can understand what will happen as we increase the coupling coefficient κ. First we shall see only spatially homogeneous oscillations. Then at the value of κ above, the regime of oscillations will switch to the first propagating wave. As we increase κ further, new periodic trajectories appear in the phase space of our system. At some point a chaotic set is formed, as has been confirmed by numerical experiment, which involves the periodic trajectories corresponding to new unstable periodic waves. Since the number of positive Lyapunov exponents of this trajectory increases as the coupling coefficient decreases, so should the Lyapunov dimension.

This also confirmed by the numerical computation reported in Figure 25.1.

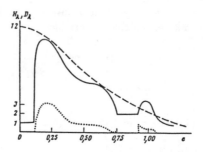

Figure 25.1

As κ becomes smaller, the dimension of the chaotic set becomes larger and the time series looks more and more like pure noise. An experimenter may run into a big problem trying to understand the dynamical nature of this "noise", if he(she) did not observe the chain of bifurcations shown in the Figure.

26 Problems to Accompany the Lectures

26.1 Lecture One

1.1 Derive the equations of motion for the damped pendulum with variable length $l(t)$. If $l(t) = l_0(1 + a\cos(\omega t))$; $a \ll 1$, find the equations of motion for the pendulum. When is the oscillation at the pendulum natural frequency stable to small perturbations?

1.2 Solve the equation $da(t)/dt = a(t)^2$ with initial condition $a(0)$. Why is this not an acceptable physical equation? How would you modify it to change this?

1.3 Using the equation $\ddot{\phi} + \omega^2 \sin(\phi) = 0$, establish the phase space diagram shown in Figure 1.4. Verify by considering infinitesimal displacements about the equilibrium points at $\phi = 0$ and $\phi = \pi$ which points are stable to small perturbations and which are unstable. Describe qualitatively what happens to small displacements near the point $\phi = \pi$.

26.2 Lecture Two

2.1 Verify that a trajectory started along a separatrix will take a logarithmically infinite amount of time to reach a fixed or equilibrium point on that separatrix.

2.2 If the phase plane has ellipses in it, such as in Figure 2.4, what can you say about the stability of orbits near the fixed point in the center?

2.3 If a dynamical system is Hamiltonian, that is the equations of motion are derivable from a scalar function $H(p, q)$ via

$$\frac{dq}{dt} = \frac{\partial H(p, q)}{\partial p}$$

$$\frac{dp}{dt} = -\frac{\partial H(p,q)}{\partial q},$$

discuss the possible patterns of linear stability around fixed points (p_0, q_0) when we have only one degree of freedom: one p and one q.

2.4 Verify in detail the statements in the text about the solutions to the equation $\ddot{x} - x + x^2 = 0$.

26.3   Lecture Three

3.1 Using Equation (10) determine what range of the parameters a,b,c,d correspond to each of the possibilities in Figure 3.1.

3.2 Consider the equations of motion for the function $z(t) = x(t) + iy(t)$

$$\frac{dz(t)}{dt} = (a - |z(t)|^2)z(t) + \epsilon. \tag{182}$$

Express $z(t)$ in polar coordinates $z(t) = r(t)e^{i\theta(t)}$ and show

$$\frac{dr(t)}{dt} = (a - r^2)r + \epsilon\cos(\theta)$$

$$\frac{d\theta(t)}{dt} = -\frac{\epsilon\sin(\theta)}{r},$$

and find the equilibrium points of this system as the parameters a and ϵ vary. Form a bifurcation diagram as in Figure 3.1.

26.4   Lecture Four

4.1 Verify the details of Equation (12).

4.2 Choose one of the physical systems shown in Figure 4.1 and demonstrate that the equations of motion can be written in the form

$$\frac{dx}{dt} = y$$

$$\frac{dy}{dt} = -x + \mu[-q(x)y + f(x)],$$

which is studied in this Lecture. Identify the small parameter μ, and tell when the approximations here will be appropriate.

4.3 In the concrete example in the Lecture $q(x) = 1 - \alpha x^2 + \beta x^4$; $f(x) = 0$. Verify the equations of motion for the modulus A of a(t)

$$\frac{dA(t)}{dt} = -\frac{\partial U(A)}{\partial A}, \tag{183}$$

with $U(A)$ as given.

26.5   Lecture Five

5.1 Using the expression for the current-voltage characteristic of a tunnel diode, derive the equations of motion for the circuit shown in Figure 5.1. Verify in detail that this can be written in the form $\ddot{x} - \mu(1 - \beta x^2)\dot{x} + \omega_0^2 x = 0$.

5.2 From the form of the Van der Pol equation and for $\mu \ll 1$, establish that the modulus $A(t)$ obeys an equation of gradient form as given in the lecture. Discuss solutions to this equation for the modulus $A(t)$.

5.3 Find a solution of the equation

$$\ddot{x} + \alpha x + \delta x^2 = \Gamma \cos(4t), \tag{184}$$

in the form $x(t) = C + A\cos(2t)$, for $\alpha, \delta, \Gamma > 0$, for certain values of α. Determine the stability of these periodic solutions.

26.6   Lecture Six

6.1 Suppose the motion of some dynamical system lies on a two dimensional torus. Describe what kind of motions would be seen on various Poincaré sections consisting of a plane which might be passed through the torus.

6.2 If we have a Hamiltonian system with two canonical degrees of freedom $\mathbf{q} = [q_1, q_2]$ and $\rho = [p_1, p_2]$, one may create a Poincaré section—indeed, the original section of Poincaré–by considering the surface of constant energy $H(\rho, \mathbf{q}) = E$ on which the motion occurs and then putting a point on the (q_1, p_1) plane with $q_2 = 0$ each time the orbit passes through the plane. This creates a two dimensional map $[q_1(n), p_1(n)] \to [q_1(n+1), p_1(n+1)]$ where the integer 'n' labels the order of passage through the plane. Prove that this evolution in the plane is a canonical evolution; that is, the evolution preserves Poisson brackets from one passage through the plane to the next. What does a periodic orbit in the full space look like on the section? What constraints are there on the stability of a fixed point in the plane?

6.3 In the map $x(n+1) = \lambda x(n)[1 - x(n)]$ show that $x(n) = 0$ is always a fixed point of the map. For what values of λ is it stable?

6.4 For the map of the plane to itself

$$\begin{aligned} x(n+1) &= x(n)\cos(\theta) - y(n)\sin(\theta) + x(n)^2 \sin(\theta) \\ y(n+1) &= x(n)\sin(\theta) + y(n)\cos(\theta) - x(n)^2 \sin(\theta), \end{aligned} \tag{185}$$

show that it is area preserving in the (x, y) plane. Find its fixed points and examine their stability.

26.7   Lecture Seven

7.1 For the conveyor belt in Figure 7.4, derive the equations of motion for this physical system. Cast them into the form in the Lecture.

7.2 Verify numerically the time evolution shown in Figure 7.3.

26.8   Lecture Eight

8.1 Write the equations of motion for the driven, damped pendulum in Figure 8.1. In the approximation that the displacement from the vertical is small, arrive at the equation in the Lecture. When the displacement is not small, what happens near resonance?

8.2 Verify the properties of the equations for $A(t)$ and $\Phi(t)$ in the Lecture. In particular, find the equation for the resonance curve.

8.3 Find a solution of the equation

$$\ddot{x} + x + ax^3 = b\cos(3\omega t), \tag{186}$$

in the form $x(t) = A\cos(\omega t)$, when $\omega^2 = 1 + 3[\frac{b^2 a}{4}]^{1/3}$. Find the stability of this solution under small perturbations.

26.9   Lecture Nine

9.1 Verify the equations for $A(t)$ and $B(t)$, Equation (18), in Lecture Nine.

9.2 Derive the stability regions in Figure 9.4 and establish that the behavior of the system in each region is as shown.

26.10   Lecture Ten

10.1 Using the equations for circuits, derive Equation (26).

10.2 Using the averaging methods of the lectures, derive Equation (37) and then (44).

10.3 Find the fixed (equilibrium) points of (44) and determine their linear stability.

26.11   Lecture Eleven

11.1 In two dimensions, show that the Poincaré index for an elliptic fixed point is $+1$. What about for a saddle–stable in one direction and unstable in another.

11.2 Verify that in the case of the Andronov-Hopf bifurcation from a fixed point to a limit cycle, the Poincaré index is unchanged.

26.12 Lecture Twelve

12.1 Derive the equations of motion for the spring-pendulum if it is allowed to move in three dimensions. Start with the Lagrangian for the problem and write the equations of motion in Hamiltonian form. Is the system integrable–can one find a solution in closed form?

12.2 How are the conserved quantities in the Manley-Rowe relations related to the action variables in the Hamiltonian description of the spring-pendulum? What does the approximate conservation mean for the appearance of orbits of the system?

12.3 In Euler's equation for a rotating rigid body, investigate the stability of the equilibrium states representing rotation about one of the axes alone. Do the same now for the amplitude-phase equations (63).

26.13 Lecture Thirteen

13.1 Derive the equations of motion and dispersion relation (70) for the linear coupled oscillators.

13.2 From the Sine-Gordon equation (72) deduce the equation for traveling waves (73) and verify that the solutions in (75) and (76) are correct.

13.3 For the KdV equation (83) verify that (84) is a solution. Is it stable? Show how you would find out.

26.14 Lecture Fourteen

14.1 Verify that the steady wave form of the KdV-Burgers equation is given by (91) and that (92) is true.

14.2 For (92) describe the phase portrait in the phase space of $u(\xi)$, $\frac{du(\xi)}{d\xi}$.

26.15 Lecture Fifteen

15.1 Starting from the two dimensional (x, y) fluid equations for a shallow fluid in uniform gravity in the y direction, derive the shallow water equations (95) and (96). Consider the x velocity $u(x, y, t)$ independent of y and use the fact that length scales in x are much longer than in y. The height of the fluid $h(x, t)$ acts as a density in the shallow water equations; explain this in physical terms. The shallow water equations supports sound waves. Find them.

15.2 Verify (104) and the statements in the lecture material after it.

26.16   Lecture Sixteen

16.1 Verify the statements made after Equation (109) about the implications of the resonance condition for the KdV-Burgers equation.

16.2 What is the damping for a mode of wave number $\mathbf{k}$ and frequency $\omega(\mathbf{k})$ for the KdV-Burgers equation?

16.3 Derive Equation (110) using the methods of earlier lectures, verify that there is an 'energy' integral.

16.4 Sketch the phase portraits for Equation (111) at various values of the detuning $\Delta\omega$.

26.17   Lecture Seventeen

17.1 For the three wave interaction in Equation (112) verify the equations for the mode intensities $N_j = |a_j(t)|^2$: Equation (114).

17.2 Show that the Manley=Rowe relations are true for these intensities.

17.3 When N_3 has reached its equilibrium state, namely one of the states in Equation (121), what is the evolution of $N_{1,2}(t)$ after that?

26.18   Lecture Eighteen

18.1 For Raleigh-Bénard convection, begin with the coupled Navier-Stokes and heat transfer equations and derive the critical curve for Rayleigh number versus wavenumber as shown in Figure 18.2. Give a qualitative explanation in terms of friction (viscosity) and box size for the two rising ends of the curve for small and for large wavenumbers.

18.2 Verify that when $\rho_{jl} = \rho_{lj}$, (127) is a gradient system. What physical principle is satisfied when the couplings are symmetric like this, and why, then, does a gradient system result?

26.19   Lecture Nineteen

19.1 How predictable is a sine wave $x(t) = A\sin(\omega t + \phi)$?

19.2 How predictable is a series of truly random numbers? How well can you approximate such a set of random numbers using some numerical algorithm?

19.3 For the Lorenz equations

$$\begin{aligned}
\dot{x}(t) &= \sigma(y(t) - x(t)) \\
\dot{y}(t) &= -y(t) + rx(t) - x(t)z(t) \\
\dot{z}(t) &= -bz(t) + x(t)y(t)
\end{aligned}$$

determine the value of r, for fixed σ and b, at which all fixed points become unstable. What happens to orbits for r greater than this value?

19.4 If $f(z) = az + bz^2 + c(e^z - 1)$, determine the stability of the fixed point at $(x, y, z) = (0, 0, 0)$. Then determine numerically what the orbits look like when all fixed points are unstable.

26.20 Lecture Twenty

20.1 In a one dimensional map $y(j+1) = f(y(j))$ verify that when $|\frac{d(y(j+1))}{dy(j)}| > 1$, the orbit is unstable.

20.2 For the map $y(j+1) = [2y(j)]$, verify this stability criterion and the statements made in the lecture.

20.3 For the same map verify in detail the statements about the invariant measure. What is the variation about the mean value for this distribution?

26.21 Lecture Twenty-one

21.1 For the logistic map $x(n+1) = kx(n)(1 - x(n))$ find the region of stability for fixed points and for period two orbits where $x(n+2) = x(n)$.

21.2 Verify the numerical statement in Equation (137) for the logistic map. What is true for the 'sine' map: $x(n+1) = k\sin(\pi x(n))$?

26.22 Lecture Twenty-two

22.1 Sketch the phase portrait for the system Equation (140). Determine the stability of any fixed points.

22.2 For (140) establish the existence of the Lyapunov function shown in (141) and verify (142).

22.3 When $\beta > 0$ find the behavior of the system (140) by numerical methods. Make a picture of the orbits in the space $[x(t), \dot{x}(t), q(t)]$.

22.4 Verify the statements about the solution of (147) made in the lecture.

26.23   Lecture Twenty-three

23.1 Numerically verify the statements made in the lecture about the circle map: $x(n+1) = x(n) + w + K\sin(x(n))$.

26.24   Lecture Twenty-four

24.1 For the Lorenz attractor whose equations have been given in the lectures and in Problem 19.3, determine the fractal dimension of the attractor using the definition of (157). The answer is $D_A \approx 2.06$.

24.2 Determine the Lyapunov exponent for the logistic map $x(n) \rightarrow kx(n)(1-x(n))$.

24.3 How much computation is required for the correlation integral defined by (166)? If the number of points is small, how well can we determine d_C?

26.25   Lecture Twenty-five

25.1 For the Ginzburg–Landau equation establish the relations (175) for the suggested solution (174).

25.2 Determine the linear stability of the discrete equation (178) for both traveling waves and the trivial solution $a = 0$.